SOTHIS PRESS

Sothis Press aims to completely reimagine the public domain book with an emphasis on universal access to captivatingly designed print editions. We believe that technology will enable every single book ever published to be perpetually accessible in beautiful print form. Sothis is beauty in ubiquity.

www.sothis.press
1 Waterhouse Square, London, ECIN 2ST

A Guide to the West

Containing Sketches of Michigan, Ohio, Indiana, Illinois, Missouri, Arkansas, Wisconsin and Ohio

J. M. Peck

SOTHIS PRESS

ISBN 978-1-0365-1440-2

A Guide to the West is
in the public domain.

A NEW GUIDE TO THE WEST:

Containing sketches of

Ohio, Michigan, Indiana, Illinois, Missouri, Arkansas, Wisconsin and Iowa.

BY J. M. PECK, A. M.,
Of Rock Spring, Ill.

CINCINNATI:
D. ANDERSON.
1848.

CONTENTS.

CHAPTER V.

PUBLIC LANDS.

CHAPTER VI.

ABORIGINES.

CHAPTER VII.

WESTERN PENNSYLVANIA.

CHAPTER VIII.

WESTERN VIRGINIA.

CHAPTER IX.

MICHIGAN.

INTRODUCTION.

MUCH has been published already about the WEST,—the GREAT WEST,—the VALLEY OF THE MISSISSIPPI;—but no portion of this immense and interesting region is so much the subject of inquiry, and so particularly excites the attention of the emigrant, as the States of Ohio, Indiana, Illinois, Missouri and Michigan, with the adjacent territorial regions.

All these States have come into existence, as such, with the exception of Ohio, within the last twenty years; and much of the territory, now adorned by the hand of civilization, and spread over with an enterprising, industrious and intelligent people,—the field of public improvements in canals and railways,—of colleges, churches, and other institutions,—was the hunting ground of the aborigines, and the scene of border warfare. These States have been unparalleled in their growth, both in the increase of population and property, and in the advance of intellectual and moral improvement. Such an extent of forest was never before cleared,—such a vast field of prairie was never before subdued and cultivated by the hand of man, in the same short period of time. Cities, and towns, and villages, and counties, and States, never before rushed into existence and made such giant strides, as upon this field.

"*Who hath heard such a thing? Who hath seen such things? Shall the earth be made to bring forth in one day? or shall a nation be born at once?*"*

* Isaiah 66: 8

The rapid increase of population will be exhibited in a tabular form, in the following pages, and other parts, showing that the general improvement of the country, and the development of its physical, intellectual and moral resources have kept pace with the extension of settlements. And such are its admirable facilities for commerce, by its numerous navigable rivers, and its lines of canals, some of which are finished, and many others commenced or projected,—such the richness of its soil and the variety of its productions,—such the genial nature of its climate,—the enterprise of its population, and the influence it must soon wield in directing the destinies of the whole United States,—as to render the GREAT WEST an object of the deepest interest to the American patriot. To the philanthropist and Christian, the character and manners, the institutions, literature and religion of so wide a portion of our country, whose mighty energies are soon to exert a controlling influence over the character of the whole nation, and in some measure of the world,—are not less matters of momentous concern.

"The West is a young empire of mind, and power, and wealth, and free institutions, rushing up to a giant manhood with a rapidity and power never before witnessed below the sun. And if she carries with her the elements of her preservation, the experiment will be glorious,—the joy of the nation,—the joy of the whole earth, as she rises in the majesty of her intelligence, and benevolence, and enterprise, for the emancipation of the world."*

Amongst the causes that have awakened the attention of the community in the Atlantic States to this great Valley, and excited the desires of multitudes to remove hither, may be reckoned the efforts of the liberal and benevolent to aid the West in the immediate supply of her population with the Bible, with Sunday schools, with religious tracts, with the gospel ministry, and to lay the foundation for colleges and other literary institutions. Hundreds of families, who might otherwise have remained in the crowded cities and densely populated neighborhoods of their ancestors, have

* Beecher.

had their attention directed to these States as a permanent home. And thousands more, of virtuous and industrious families, would follow, and fix their future residence on our prairies, and in our western forests, cultivate our wild lands, aid in building up our towns and cities, and diffuse a healthful, moral and intellectual influence through the mass of our present population, could they feel assured that they can reach some portion of the Western Valley without great risk and expense, provide for their families comfortably, and not be swept off by sickness, or overwhelmed by suffering beyond what is incident to any new country.

The author's first book, 'A GUIDE FOR EMIGRANTS,' &c., was written in the winter and spring of 1831, to answer the pressing call then made for information of these Western States, but more especially that of Illinois;—but many of its particulars, as to the character and usages of the people, manners and customs, modes of erecting buildings, general characteristics and qualities of soil, productions, &c., were applicable to the West generally.

Since that period, brief as it has been, wide and rapid changes have been made, population has rapidly augmented, beyond that of any former period of the same extent; millions of acres of the public domain, then wild and hardly explored, have been brought into market; settlements and counties have been formed, and populous towns have sprung up, where, at that time, the Indian and wild beast had possession; facilities for intercommunication have been greatly extended, and distant places have been brought comparatively near; the desire to emigrate to the West has increased, and every body in the Atlantic States has become interested, and inquires about the great Valley. That respectable place, so much the theme of declamation and inquiry abroad, "*the Far West*," has gone from this region towards the setting sun. Its exact locality has not yet been settled, but probably it may soon be found along the Gulf of California, or near Nootka Sound. And, if distance is to be measured by time and the facility of intercourse, we are now several hundred miles nearer the Atlantic coast than twenty years since. Ten years more, and the facilities of railways and improved machinery will place the Mississippi within seven days' travel of Boston, six days of Washington city, and five days of Charleston, S. C.

To give a brief, and yet correct account of a portion of this great Valley, its resources, the manners and customs of its inhabitants, its political subdivisions, cities, commercial and other important towns, colleges and other literary institutions, religious condition, public lands, qualities of soil and general features of each State and Territory named in the title page, together with such information as may form a kind of manual for the emigrant and man of business, or which may aid him on his journey hither, and enable him to surmount, successfully, the difficulties of a new country, is the object of this new work. In accomplishing this task the author has aimed at *correctness* and *brevity*. To condense the particular kind of information called for by the public mind, in a small space, has been no easy task. Nor has it been a small matter to collect from so wide a range as five large States, and two extenive Territories, with other large districts, the facts and statistical information often found in the compass of less than a page.

It is an easy task to a belles-lettres scholar, sitting at his desk, in an easy chair, and by a pleasant fire, to write "Histories," and "Geographies," and "Sketches," and "Recollections," and "Views," and "Tours," of the Western Valley,—but it is quite another concern to explore these regions, examine public documents, reconcile contradictory statements, correspond with hundreds of persons in public and private life, read all the histories, geographies, tours, sketches and recollections, that have been published, and correct their numerous errors,—then collate, arrange, digest, and condense the facts of the country. Those who have read his former "Guide for Emigrants," will find upon perusal, that this is radically a *new work*—rather than a new edition. Its whole plan is changed; and though some whole pages of the former work are retained, and many of its facts and particulars given in a more condensed form, much of that work being before the public in other forms, he has been directed, both by his own judgment, and the solicitude of the public mind in the Atlantic States, to give to the work its present form and features.

There are three classes of persons in particular, who may derive advantage from this Guide.

1. All those who intend to remove to the States and Territories described. Such persons, whether citizens of the Atlantic States, or natives of Europe, will find in this small volume much of that species of information for which they are solicitous.

It has been a primary object of the author throughout this work, to furnish the outline of facts necessary for this class. He is aware also, that much in detail will be desired and eagerly sought after, which the portable and limited size of this little work could not contain; but such information may be found in the larger works of Hall, Flint, Darby, Schoolcraft, Long, and other authors and travelers. Those who desire more specific and detailed descriptions of Illinois, will be satisfied, probably, with the author's GAZETTEER of that State, published in 1834, and which can be had on application to the author, or to the publishers of this work.

2. This Guide is also designed for those, who, for either pleasure, health or business, intend to travel through the Western States. Such are now the facilities of intercommunication between the Eastern and Western States, and to most points in the Valley of the Mississippi, that thousands are visiting some portions of this interesting region every month. Some knowledge of the routes that lead to different parts of this Valley, the lines of steam-boats and stages, cities, towns, public institutions, manners and customs of the people, &c., is certainly desirable to all who travel. Such persons may expect a correct, and it is hoped a pleasant Guide in this book.

3. There is a numerous class of persons in the Atlantic States, who desire to know more about the Great West, and to have a book for reference, who do not expect to emigrate here. Many are deeply interested in its moral welfare. They have cheerfully contributed to establish and build up its literary and religious institutions, and yet from want of access to those facts which exist amongst us, their information is but partial and limited. The author, in his travels in the Atlantic States, has met with many persons, who, though well informed on other subjects, are

surprisingly ignorant of the actual condition, resources, society, manners of the people, and even the geography of these States and Territories. The author is aware of the difficulty of conveying entirely correct ideas of this region, to a person who has never traveled beyond the borders of his native State. The laws and habits of associating ideas in the human mind forbid it.

The chief source of information for those States that lie on the Mississippi, has been the personal observation of the author,—having explored most of the settlements in Missouri and Illinois, and a portion of Indiana and Ohio,—having spent more than eighteen years here, and seen the two former States, from an incipient Territorial form of government, and a few scattered and detached settlements, arise to their present state of improvement, population, wealth and national importance. His next source of information has been from personal acquaintance and correspondence with many intelligent citizens of the States and Territories he describes. Reference has also been had to the works of Hall, Flint, Darby, Breckenridge, Beck, Long, Schoolcraft, Lewis and Clarke, Mitchell's and Tanner's Maps, Farmer's Map of Michigan, Turnbull's Map of Ohio, Ohio Gazetteer, Indiana Gazetteer, Dr. Drake's writings, M'Coy's Annual Register of Indian Affairs, Ellicott's Surveys, and several periodicals. J. M. P.

Rock Spring, Illinois, January, 1836.

GUIDE TO EMIGRANTS.

CHAPTER I.

GENERAL VIEW OF THE VALLEY OF THE MISSISSIPPI.

Its Extent—Subdivisions—Population—Physical Features —Rivers.

THE Valley of the Mississippi, in its proper geographical extent, embraces all that portion of the United States, lying between the Alleghany and Rocky Mountains, the waters of which are discharged into the gulf of Mexico, through the mouths of the Mississippi. I have embraced, however, under that general term, a portion of the country bordering on the northern lakes, including the north part of Ohio, the north-eastern portions of Indiana and Illinois, the whole of Michigan, with a considerable territorial district on the west side of lake Michigan, and around lake Superior.

Extent. This great Valley is one of the

largest divisions of the globe, the waters of which pass one estuary.

To suppose the United States and its territory to be divided into three portions, the arrangement would be, the Atlantic slope,—the Mississippi basin, or valley,—and the Pacific slope.

A glance on any map of North America, will show that this Valley includes about two-thirds of the territory of the United States. The Atlantic slope contains about 390,000; the Pacific slope, about 300,000; which, combined, are 690,000 square miles; while the Valley of the Mississippi contains, at least, 1,300,000 square miles, or 833,000,000 acres.

This Valley extends from the 29° to the 49° of north latitude, or about 1400 miles from south to north; and from the 3° to the 35° of longitude west from Washington, or about 1470 miles from east to west. From the source of the Alleghany river to the sources of the Missouri, following the meanderings of the streams, is not less than 5000 miles.

Subdivisions. The States and Territories included, are a small section of New York, watered by the heads of the Alleghany river, Western Pennsylvania, Western Virginia, Ohio, Indiana, Illinois, Missouri, Michigan, Kentucky, Tennessee, Mississippi, Louisiana, Arkansas, Indian Territory, the vast unsettled regions lying to the west and north of this Territory, the Wisconsin Territory, including an extensive country west of the Mississippi, and north of the State of Missouri, with the vast regions that lie towards the heads of the Mississippi, and around lake Superior.*

Population. Probably, there is no portion of the globe, of equal extent, that contains as much of soil fit for cultivation, and which is capable of sustaining and supplying with all the necessaries and conveniences, and most of the luxuries of life, so dense a population as this great Valley. Deducting one third of its surface, for water and desert, which is a very liberal allowance, and there remains 866,667 square miles, or 554,666,880 acres of arable land.

The following table, gives a comparative view of the population of the Valley of the Mississippi, and shows the proportional increase of the several States, parts of States, and Territories, from 1790 to the close of 1835, a period of forty-five years. The column for 1835, is made up, partly, from the census, taken in several States and Territories, and partly by estimation. It is sufficiently accurate, for general purposes:

* Why the names, Huron, Mandan, Sioux, Osage, and *Ozark* have been applied by Darby, and other authors, to the extensive regions on the Upper Mississippi, the Upper Missouri, and the Arkansas rivers, I am not able to solve. *Osage* is a French corruption of *Wos-sosh-e*, and *Ozark* is an awkward, illiterate corruption of Au Kauzau. *Sioux* is another French corruption, the origin of which is not now easily ascertained. Carver, and other travellers, call this nation of Indians, Nau-do-wes-sees. Chiefs of this nation have repeatedly disclaimed the name of Sioux (pronounced Soos). They sometimes call themselves Da-co-tah.

States, parts of States, and Territories.	1790	1800	1810	1820	1830 *	1835
Western Pennsylvania, and a fraction of New York,	75,000	130,000	240,000	290,000	380,000	490,000
Western Virginia,	45,000	75,000	100,000	147,178	204,175	230,000
Ohio,		a45,000	230,760	581,434	937,679	1,375,000
Indiana,			24,520	147,178	341,582	600,000
Illinois,			12,282	55,211	157,575	272,427
Missouri,			b 20,845	66,586	140,074	210,000
Michigan,			4,762	8,896	31,000	83,000
Kentucky,	73,677	220,959	406,511	564,317	688,844	748,844
Tennessee,	35,691	105,602	261,727	422,813	684,822	735,000
Mississippi,		c 8,850	40,352	75,448	136,806	300,000
Louisiana,			76,556	153,407	214,693	270,000
Arkansas,				14,273	30,608	51,809
*Wisconsin Territory, and New Purchase,					d 3,608	30,000
Total,	229,368	585,411	1,418,315	2,526,741	3,951,466	5,396,080

a. Including Indiana, Illinois and Michigan.
b. Including Arkansas
c. Including Alabama.
d. Included with Michigan, in the census of 1830.
* The country west of the Mississippi, and north of the State of Missouri, was ceded by the Sauk Indians, Sept. 1832. It now contains about 10,000 inhabitants.

Let this Valley become as populous as Massachusetts, which contains 610,014 inhabitants, on an area of 7800 square miles, or seventy-eight to every 640 acres, and the population of this immense region will amount to 67,600,000. The child is now born which will live to see this result. Suppose its population to become equally dense with England, including Wales, which contains 207 to the square mile, and its numbers will amount to 179,400,000. But let it become equal to the Netherlands, the most populous country on the globe, containing 230 to the square mile, and the Valley of the Mississippi teems with a population of two hundred millions,—a result which may be had in the same time that New England has been gathering its two millions. What reflections ought this view to present to the patriot, the philanthropist, and the Christian!

Physical Features. The physical features of this Valley are peculiar.

1. It includes two great inclined planes, one on its eastern, and the other on its western border, terminating with the Mississippi.

2. This river receives all the waters produced on these slopes, which are discharged by its mouths into the gulf of Mexico.

3. Every part of this vast region can be penetrated by steam-boats, or other water craft; nor is there a spot in all this wide region, excepting a small district in the vast plains of Upper Missouri, that is more than one hundred miles from some navigable water.

A boat may take in its lading on the banks of the Chatauque lake, in the State of New York; another may receive its cargo in the interior of Virginia; a third may start from the rice lakes, at the head of the Mississippi; and a fourth may come, laden with furs, from the Chippewan mountains, 2800 miles up the Missouri, and all meet at the mouth of the Ohio, and proceed in company to the ocean.

4. With the exception of its eastern and western borders, there are no mountains. Some portions are level; a large part is gently undulating, or what in the West is called "rolling;" and the remainder is made up of abrupt hills, flint and limestone ridges, bluffs and ravines.

5. It is divided into two great portions,—the Upper and Lower Valley, according to its general features, climate, staple productions, and habits of its population. The parallel of latitude that cuts the mouth of the Ohio river, will designate these portions with sufficient accuracy.

North of this line, the seasons are regularly divided into spring, summer, autumn, and winter. In the winter there is usually more or less snow: ice forms, and frequently blocks up the rivers, and navigation is obstructed. Cotton is not produced in sufficient quantity or quality to make it a staple for exportation. It is the region of furs, minerals, tobacco, hemp, live stock, and every description of grain and fruit that grows in New England.

Its white population are mostly accustomed to labor.

South of this line, cotton, tobacco, indigo, and sugar, are staples. It has little winter; snow seldom covers the earth; ice never obstructs the rivers; and most of the labor is done by slaves.

Rivers. The rivers are the Mississippi and its tributaries; or, more correctly, the Missouri, and its tributaries. If we except the Amazon, no river can compare with this, for length of its course, the number and extent of its tributaries, the vast country they drain, and their capabilities for navigation. Its tributaries generally issue either from the eastern or western mountains, and flow over this immense region, diffusing not only fertility to the soil, but affording facilities for commerce a great part of the year.

The Missouri is unquestionably the main stream, for it is not only longer, and discharges a larger volume of water, than the Mississippi, above its mouth, but it has branches, which, for the extent of country they drain, their length, and the volume of water they discharge, far exceed the Upper Mississippi.

The characteristics of these two rivers are each distinctly marked. The Missouri is turbid, violent in its motions, changing its currents; its navigation is interrupted, or made difficult, by snags, sawyers, and planters; and it has many islands and sand-bars. Such is the character of the Mississippi, below the mouth of the Missouri. But above its mouth, its waters are clear, its current gentle, while it is comparatively free from snags and sand-bars.

The Missouri, which we have shown to be the principal stream, rises in the Chippewan, or Rocky mountains, in latitude 44° north, and longitude about 35° west from Washington city. It runs a north-east course, till after it receives the Yellow Stone, when it reaches past the 48th degree of latitude; thence an east, then a south, and, finally, a south-eastern course, until it meets the current of the Mississippi, twenty miles above St. Louis, and in latitude 38° 45′ north. Besides numerous smaller streams, the Missouri receives the Yellow Stone and Platte, which of themselves, in any other part of the world, would be called large rivers, together with the Sioux, Kausau, Grand, Chariton, Osage, and Gasconade, all large and navigable rivers.

Its length, upon an entire comparative course, is 1870 miles, and, upon a particular course, about 3000 miles. Lewis and Clark make the distance from the Mississippi to the great falls, 2530 miles.

There are several things in some respects peculiar to this river, which deserve notice:

1. Its current is very rapid, usually at the rate of four or five miles an hour, when at its height; and it requires a strong wind to propel a boat, with a sail, against it. Steam over-

comes its force, for boats ply regularly from St. Louis to the towns and landings on its banks, within the borders of the State, and return with the produce of the country. Small steam-boats have gone to the Yellow Stone for furs.

Owing to the shifting of its current, and its snags and sand-bars, its navigation is less safe and pleasant than any other western river, but these difficulties are every year lessened by genius and enterprise.

2. Its water is always turbid, being of a muddy ash color, though more so at its periodical rise than at other times. This is caused by extremely fine sand, received from the neighborhood of the Yellow Stone. During the summer flood, a tumbler of water taken from the Missouri, and precipitated, will produce about one fourth of its bulk in sediment.

This sediment does not prevent its habitual use by hundreds who live on its banks, or move in boats over its surface. Some filtrate it, but many more drink it, and use it for culinary purposes, in its natural state.

When entirely filtrated, it is the most limpid and agreeable river water I ever saw. Its specific gravity is then about equal to rain water; but, in its turbid state, it is much heavier than ordinary river water; for a boat will draw three or four inches less in it than in other rivers, with the same lading; and the human body will swim in it with but very little effort.

It possesses some medicinal properties. Placed in an open vessel, and exposed to the summer's sun, it remains pure for weeks. Eruptions on the skin and ulcerous sores are cured by wading or frequent bathings; and it commonly produces slight cathartic effects upon strangers, upon its first use.

The width of the Missouri river, at St. Charles, is 550 yards. Its alluvial banks, however, are insecure, and are not unfrequently washed away, for many yards, at its annual floods. The bed of its channel is also precarious, and is elevated, or depressed, by the deposition or removal of its sandy foundation. Hence, the elevation or depression of the surface of this river affords no criterion of its depth, or of the volume of water it discharges at any one period.

Undulatory motions, like the boiling of a pot, are frequently seen on its surface, caused by the shifting of the sand that forms its bed.

The volume of water it ordinarily discharges into the Mississippi, is vastly disproportionate to its length, or the number and size of its tributaries. I have seen less than six feet depth of water at St. Charles, at a low stage, and it was once forded by a soldier, at Belle-Fontaine, four miles above its junction with the Mississippi.

Evaporation takes up large quantities, but absorption, throughout the porous soil of its wide bottoms, consumes much more. In all the wells dug in the bottom lands of the Missouri, water is always found at the depth of the surface of the river, and invariably rises or sinks with the floods and ebbings of the stream. Sand frequently enters these wells as the river rises.

Its periodical floods deserve notice. Ordinarily, this river has three periods of rising and falling, each year. The first rise is caused by the breaking up of winter on the Gasconade, Osage, Kausau, Chariton, Grand, and other branches of the Lower Missouri, and occurs the latter part of February, or early in March. Its second rise is usually in April, when the Platte, Yellow Stone, and other streams pour into it their spring floods. But the flood, that more usually attracts attention, takes place from the 10th to the 25th of June, when the melting snows on the Chippewan mountains pour their contents into the Missouri. This flood is scarcely ever less than five, nor more than twenty feet, at St. Louis, above the ordinary height of the river. On two occasions, however, since the country was known to the French, it has risen to that height in the Mississippi, as to flow over the American bottom, in Illinois, and drive the inhabitants of Cahokia and Kaskaskia from their villages to the bluffs. Rain, in greater or less quantities, usually falls during the rise of the river, and ceases when the waters subside. So uniform is this the case in Upper Missouri, the region beyond the boundary of the State, that the seasons are divided into wet and dry.

Pumice stone, and other volcanic productions, occasionally float down its waters.

Mississippi River. The extreme head of the longest branch of the Mississippi river has been found in lake Itaska, or Lac la Biche, by Mr. Schoolcraft, who states it to be elevated 1500 feet above the Atlantic ocean, and distant 3160 miles from the extreme outlet of the river, at the gulf of Mexico. The outlet of Itaska lake, which is connected with a string of small lakes, is ten or twelve feet broad, and twelve or fifteen inches deep. This is in latitude about 48° north. From this, it passes Cedar, and several smaller lakes, and runs a winding course, 700 miles, to the falls of St. Anthony, where its waters are precipitated over a cataract of sixteen or seventeen feet, perpendicular. It then continues a south-eastern course, to the Missouri, in latitude 38° 38′ north, receiving the St. Croix, Chippewa, Wisconsin, Rock, and Illinois rivers, with many smaller streams from the east, and the St. Peter's Iowa, Des Moines and Salt rivers, besides a number of smaller ones from the west. The current of the Missouri strikes that of the Mississippi, at right angles, and throws it upon the eastern shore. When at a low stage, the waters of the two rivers are distinct till they pass St. Louis.

The principal branch of the Upper Mississippi is the St. Peter's, which rises in the great prairies in the north-west, and enters the parent stream ten miles below the falls of

St. Anthony. Towards the sources of this river the quarries exist from which are made the red stone pipes of the Indians. This is sacred ground: hostile tribes meet here, and part unmolested.

Rock river drains the waters from the northern part of Illinois and Wisconsin, and enters the parent stream, at latitude 41° 30′ north. In latitude 39° comes in the Illinois (signifying the "River of Men"); and, eighteen miles below this, it unites with, and is lost in the Missouri.

Custom has fixed, unalterably, the name *Mississippi* to this united body of waters, that rolls its turbid waves towards the Mexican gulf; though, as has been intimated, it is but a continuation of the Missouri.

Sixty miles below St. Louis, the Kaskaskia joins it, after a devious course of 400 miles. In latitude 37° north, the Ohio pours in its tribute (called by the early French explorers, "La Belle Rivière,"—*the beautiful river*). A little below 34° the White river enters, after a course of more than 1000 miles. Thirty miles below that, the Arkansas, bringing its tribute from the confines of Mexico, pours in its waters. Above Natchez, the Yazoo, from the east, and, eighty miles below, the Red river, from the west, unite their waters with the Mississippi. Red river takes its rise in the Mexican dominions, and runs a course of more than 2000 miles.

Hitherto, the waters in the wide regions

of the West have been congregating to one point. The "Father of Waters" is now upwards of a mile in width, and several fathoms deep. During its annual floods, it overflows its banks, below the mouth of the Ohio, and penetrates the numerous bayous, lakes and swamps, and especially on its western side. In many places, these floods extend thirty or forty miles into the interior. But after it receives the Red river, it begins to throw off its surplus waters, which flow in separate channels to the gulf, and never again unite with the parent stream. Several of these communications are held with the ocean, at different and distant points.

Ohio River. The Ohio river is formed by the junction of the Alleghany and Monongahela, at Pittsburgh. The Alleghany river rises not far from the head of the western branch of the Susquehannah, in the highlands of McKean county, Pennsylvania. It runs north, till it penetrates Cataraugus county, New York; then turns west, then south-west, and finally takes a southern course, to Pittsburgh. It receives a branch from the Chatauque lake, Chatauque county, New York. The Monongahela rises near the sources of the Kenawha, in Western Virginia, and runs north till it meets the Alleghany.

The general course of the Ohio is south-west. Its current is gentle, and it receives a number of tributaries, which are noticed in the States where they run.

The Valley of the Mississippi has been arranged, by Mr. Darby, into four great subdivisions:

1. The *Ohio Valley*,—length, 750 miles, and mean width, 261; containing 196,000 square miles.

2. *Mississippi Valley*, above Ohio, including the minor valley of Illinois, but exclusive of Missouri, 650 miles long, and 277 mean width, and containing 180,000 square miles.

3. *Lower Valley of the Mississippi*, including White, Arkansas, and Red river valleys, 1000 miles long, and 200 wide, containing 200,000 square miles.

4. *Missouri Proper*, including Osage, Kausau, Platte rivers, &c., 1200 miles long, and 437 wide, containing 523,000 square miles.

"The *Valley of the Ohio* is better known than any of the others; has much fertile land, and much that is sterile, or unfit for cultivation, on account of its unevenness. It is divided into two unequal portions, by the Ohio river; leaving on the right, or north-west side, 80,000, and on the left, or south-east side, 116,000 square miles. The eastern part of this valley is hilly, and rapidly acclivous towards the Appalachian mountains. Indeed, its high hills, as you approach these mountains, are of a strongly marked mountainous character. Of course, the rivers which flow into the Ohio,—the Monongahela, Kenawha, Licking, Sandy, Kentucky, Green, Cumberland and Tennessee,—are rapid, and abounding in cataracts and falls, which, towards their sources, greatly impede navigation. The western side of this Valley is also hilly, for a considerable distance from the Ohio, but, towards its western limit, it subsides to a remarkably level region; so that, whilst the eastern line of this Valley lies along the high table land, on which the Appalachian mountains rest, and where the rivers of the eastern section of this Valley rise (which is at least 2000 miles, generally, above the ocean level), the western line has not an elevation of much more than half of that amount on the north, and which greatly subsides towards the Kaskaskia. The rivers of the western section are Beaver, Muskingum, Hockhocking, Scioto, Miama, and Wabash. Along the Ohio, on each side, are high hills, often intersected with deep ravines, and sometimes openings of considerable extent, and well known by the appellation of 'Ohio Hills.' Towards the mouth of the Ohio, these hills almost wholly disappear, and extensive level bottoms, covered with heavy forests of oak, sycamore, elm, poplar and cotton-wood, stretch along each side of the river. On the lower section of the river, the water, at the time of the spring floods, often overflows these bottoms to a great extent. This fine Valley embraces considerably more than one half of the whole population of the entire Valley of the West. The western parts of Pennsylvania and Virginia, the entire States of Ohio, Indiana and Kentucky, the larger part of Tennessee, and a smaller part of Illinois, are in the Valley of the Ohio."

The *Upper Valley of the Mississippi* possesses a surface far less diversified than the Valley of the Ohio. The country, where its most northern branches take their rise, is elevated table land, abounding with marshes and lakes, that are filled with a graniferous vegetable, called wild rice. It is a slim, shriveled grain, of a brownish hue, and gathered by the Indians, in large quantities, for food. There are tracts of arable land, covered with elm, linden, pine, hemlock, cherry, maple, birch, and other timber common to a northern climate. From the same plateau flow the numerous branches of Red river, and other streams that flow into lake Winnipec, and thence into Hudson's bay. Here, too, are found some of the head branches of the waters of St. Lawrence, that enter the Lake of the Woods and Superior. In the whole country of which we are speaking, there is nothing that deserves the name of mountain. Below the falls of St. Anthony the river bluffs are often abrupt, wild and romantic; and at their base and along the streams are thousands of quartz crystals, cornelians, and other precious stones.

But a short distance in the rear, you enter upon table land of extensive prairies, with clumps of trees, and groves, along the streams. Further down, abrupt cliffs, and overhanging precipices, are frequently seen at the termination of the river alluvion.

The whole country north-west of the Ohio, and east of the Mississippi, as far north as the falls of St. Anthony, exhibit striking marks of a diluvial formation, by a gradual retiring of the waters. From the summit level that divides the waters of the lakes from those of the Mississippi, through Ohio, Indiana, Illinois, and Wisconsin, which is scarcely a perceptible ridge, to the south point of Illinois, at the junction of the Ohio and Mississippi, appears to have once been a plane, with an inclination equal to twelve or fifteen inches per mile. The ravines and valleys appear to have been gradually scooped out by the abrasion of the waters.

"The *Lower Mississippi Valley* has a length of 1200 miles, from north-west to south-east, considering the source of the Arkansas and the mouth of the Mississippi river as extreme points; reaching from latitude 29° to 42° north, and, without estimating mountains, ridges or peaks, differs in relative elevation at least 500 feet.

"The *Arkansas river* rises near latitude 42° north, and longitude 32° west from Washington, and falls into the Mississippi at 33° 56′, passing over eight degrees of latitude.

"*Red river* rises in the mountainous country of Mexico, north of Texas, in latiude 34° north, and longitude 29° west from Washington, and falls into the Mississippi in latitude 31°. They are both remarkable rivers, for their extent, the number of their branches, the volume of their waters, the quantity of alluvion they carry down to the parent stream, and the color of their waters. Impregnated by saline particles, and colored with ocherous earth, the waters of these two rivers are at once brackish and nauseous to the taste, particularly near their mouths: that of Red river is so much so at Natchitoches, at low water, that it cannot be used for culinary purposes.

"At a short distance below the mouth of the Red river, a large bayou (as it is called), or outlet, breaks from the Mississippi, on the west; by which, it is believed, as large a volume of water as the Red river brings to the parent river, is drained off, and runs to the gulf of Mexico, fifty miles from the mouth of the Mississippi. The name of this bayou is Atchafalaya, or, as it is commonly called, *Chaffalio.* Below this bayou, another of large dimensions breaks forth on the same side, and finally falls into the Atchafalaya. This is the Placquemine. Still lower, at Donaldsonville, ninety miles above New Orleans, on the same side, the Lafourche bayou breaks out, and pursues a course parallel to the Mississippi, fifty miles west of the mouth of that river. On the east side, the Ibberville bayou drains off a portion of the waters of the Mississippi, into lakes Maurepas, Ponchartrain, Borgnes, and the gulf of Mexico, and thus forms the long and narrow island of Orleans.

"In the Lower Valley of the Mississippi there is a great extent of land of the very

2*

richest kind. There is also much that is almost always overflown with waters, and is a perpetual swamp. There are extensive prairies in this Valley; and towards the Rocky mountains, on the upper waters of the Arkansas and Red rivers, there are vast barren steppes, or plains of sand, dreary and barren, like the central steppes of Asia. On the east of the Mississippi are extensive regions of the densest forests, which form a striking contrast with the prairies which stretch on the west of that great river.

"The *Valley of the Missouri* extends 1200 miles in length, and 700 in width, and embraces 253,000 square miles. The Missouri river rises in the Chippewan mountains, through eight degrees, or nearly 600 miles. The Yellow Stone is its longest branch. The course of the Missouri, after leaving the Rocky mountains, is generally south-east, until it unites with the Mississippi. The principal branches flow from the south-west. They are the Osage, Kausau, Platte, &c. The three most striking features of this Valley are, 1st. The turbid character of its waters. 2d. The very unequal volumes of the right and left confluences. 3d. The immense predominance of the open prairies, over the forests which line the rivers. The western part of this Valley rises to an elevation towards the Chippewan mountains, equal to ten degrees of temperature. Ascending from the lower verge of this widely extended plain, wood becomes more and more scarce, until one naked surface spreads on all sides. Even the ridges and chains of the Chippewan partake of these traits of desolation. The traveler, who has read the descriptions of Central Asia, by Tooke or Pallas, will feel on the higher branches of the Missouri, a resemblance, at once striking and appalling; and he will acknowledge, if near to the Chippewan mountains, in winter, that the utmost intensity of frost over Siberia and Mongolia has its full counterpart in North America, on similar, if not on lower latitudes. There is much fertile land in the Valley of the Missouri, though much of it must be for ever the abode of the buffalo and the elk, the wolf and the deer "*

* Darby.

CHAPTER II.

GENERAL VIEW OF THE VALLEY OF THE MISSISSIPPI.

(CONTINUED.)

Productions.

Minerals. But few mines exist in the Lower Valley of the Mississippi. *Louisiana* being chiefly alluvion, furnishes only two specimens, sulphuret of antimony, and meteoric iron ore. It is supposed that the pine barrens towards Texas, if explored, would add to the number.

The only minerals in *Mississippi* are,—amethyst, of which one crystal has been found; potter's clay, at the Chickasaw bluffs, and near Natchez; sulphuret of lead, in small quantities, about Port Gibson; and sulphate of iron. Petrified trunks of trees are found in the bed of the Mississippi, opposite Natchez. In Arkansas, are various species. Here may be found the native magnet, or magnetic oxide of iron, possessing strong magnetic power. Iron ores are very abundant. Sulphate of copper, sulphuret of zinc, alum, and aluminous slate are found about the cove of Washitau and the hot springs. Buhr stone, of a superior quality, exists in the surrounding hills. The hot springs are interesting, on account of the minerals around them, the heat of their waters, and as furnishing a retreat to valetudinarians from the sickly regions ot the South. They are situated on the Washitau, a large stream that empties into Red river.

The *lead mines* of Missouri have been worked for more than a century. They are distributed through the country, from thirty to one hundred miles south-west from St. Louis, and probably extend through the Gasconade country. Immense quantities of iron ore exist in this region. Lead is found in vast quantities, in the northern part of Illinois, the south part of Wisconsin Territory, and the country on the west side of the Mississippi. These mines are worked extensively. Native copper, in large quantities, is found in the same region. Large quantities of iron ore is found in the mountainous parts of Tennessee and Kentucky, where furnaces and forges have been erected; also, in the hilly parts of Ohio, particularly at the falls of Licking, four miles west of Zanesville; and in Adams and Lawrence counties, near the Ohio river. With *iron ore* the West is profusely supplied.

Bituminous coal exists, in great profusion, in various parts of the Western Valley. The hills, around Pittsburgh, are inexhaustible. It extends through many portions of Ohio and Indiana. Nearly every county in Illinois is supplied with this valuable article. Missouri, Kentucky, and Tennessee have their share. Immense quantities are found in the mountains along the Kenawha, in Western Virginia, and it is now employed in the manufacture of salt. The Cumberland mountains, in Tennessee, contain immense deposits.

Muriate of soda, or common salt, exists in most of the States and Territories of this Valley. Near the sources of the Arkansas, incrustations are formed by evaporation, during the dry season, in the depressed portions of the immense prairies of that region. The celebrated salt rock is on the red fork of the Canadian, a branch of the Arkansas river. Jefferson lake has its water strongly impregnated with salt, and is of a bright red color. Beds of rock salt are in the mountains of this region. Several counties of Missouri have abundant salt springs. Considerable quantities of salt are manufactured in Jackson, Gallatin, and Vermilion counties, Illinois. Saline springs, and "licks," as they are called, abound through Kentucky, Tennessee, Indiana, Ohio, Western Pennsylvania, and Western Virginia. Salt is manufactured, in great abundance, at the Kenawha salines, sixteen miles above Charlestown, Va., and brought down the Kenawha river, and carried to all the Western States. Much salt is made also on the Kiskiminitas, a branch of the Alleghany river, at the Yellow creek, above Steubenville, and in the Scioto country, in Ohio. The water is frequently obtained by boring through rock, of different strata, several hundred feet deep.

Copper, antimony, manganese, and several other minerals, are found in different parts of the West, but are not yet worked. *Nitrate of potash* is found in great abundance, in the caverns of Kentucky and Tennessee, also in Missouri, from which large quantities of saltpetre are manufactured. *Sulphate of magnesia* is found in Kentucky, Indiana, and perhaps other States. Sulphur, and other mineral springs, are very common in the Western States.

Vegetable Productions, Trees, &c. Almost every species of timber and shrub, common to the Atlantic States, is found in some part of the Western Valley. The cotton-wood and sycamore are found along all the rivers below the 41st deg. of north latitude. The cypress begins near the mouth of the Ohio, and spreads through the alluvion portions of the Lower Valley. The magnolia, with its large, beautiful flower, grows in Louisiana, and the long leaf pine flourishes in the uplands of the same region. The sugar-maple abounds in the northern and middle portions. The chestnut is found in the eastern portion of the Valley, as far as Indiana, but not a tree is known to exist in a natural state, west of the Wabash river. Yellow or pitch pine grows in several counties of Missouri, especially on the Gasconade, from whence large quantities of lumber are brought to St. Louis. White pine, from the Alleghany river, is annually sent to all the towns on the Ohio, and further down. Considerable quantities of white pine grow on the Upper Mississippi, along the western shore of Michigan, about Green Bay, and along the shores of lake Superior. The yellow poplar (Liriodendron tulipifera) is a majestic tree, valuable for light boards, and may be found in some parts of most of the Western States. The beech tree is frequently found in company. The live-oak, so valuable in ship-building, is found south of the 31st degree, and along the Louisiana coast. The orange, fig, olive, pine-apple, &c., find a genial climate about New Orleans. High in the north, we have the birch, hemlock, fir, and other trees peculiar to a cold region. Amongst our fruit-bearing trees, we may enumerate the walnut, hickory, or shag-bark, persimmon, pecaun, mulberry, crab-apple, paupau, wild plum, and wild cherry. The vine grows every where. Of the various species of oak, elm, ash, linden, hackberry, &c., it is unnecessary to speak. Where forests abound, the trees are tall and majestic. In the prairie country, the timber is usually found on the streams, or in detached groves.

In the early settlement of Kentucky, there were found, south of Green river, large tracts, with stunted, scattering trees, intermixed with hazel and brushwood. From this appearance it was inferred that the soil was of inferior quality, and these tracts were denominated "barrens." Subsequently, it was found that this land was of prime quality. The term "barrens" is now applied, extensively, in the West, to the same description of country. It distinguishes an intermediate grade, from forest and prairie. A common error has prevailed abroad, that our prairie land is wet. *Prairie* is a French word, signifying *meadow*, and is applied to any description of surface, that is destitute of timber and brushwood, and clothed with grass. Wet, dry, level, and undulating, are terms of description, merely, and apply to prairies in the same sense as they do to forests. The prairies, in summer, are clothed with grass, herbage and flowers; exhibit a delightful prospect, and furnish most abundant and luxuriant pasturage for stock. Much of the forest land, in the Western Valley produces a fine range for domestic animals and swine. Thousands are raised, and the emigrant grows wealthy, from the bounties of nature, with but little labor.

Of *animals*, *birds*, and *reptiles*, little need be said. The buffalo was in Illinois the beginning of the present century. They are not found now, within three hundred miles of Missouri and Arkansas, and they are fast receding. Deer are found still in all frontier settlements. Wolves, foxes, racoons, wild cats, opossums and squirrels are plenty. The brown bear is still hunted in some parts of the Western States. Col. Crockett was a famous bear-hunter, in Western Tennessee. The white bear, mountain sheep, antelope and beaver, are found in the defiles of the Rocky mountains. The elk is still found by the hunter contiguous to newly formed settlements. All the domestic animals of the United States flourish here.

Nearly all the feathered tribe of the Atlantic slope are to be found in the Valley. Pelicans, wild geese, swans, cranes, ducks, paroquets, wild turkies, prairie hens, &c., are found in different States, especially on the Mississippi.

Reptiles. The rattlesnake, copperhead snake, moccasin snake, bull snake, and the various snakes usually found in the Atlantic States, are here. Of the venomous kinds, multitudes are destroyed by the deer and swine. Chameleons and scorpions exist in the Lower Valley, and lizards, every where. The alligator, an unwieldy and bulky animal, is found in the rivers and lakes south of latitude 34° north. He sometimes destroys calves and pigs, and, very rarely, even young children.

History. The honor of the discovery of this country is disputed by the Spanish, English and French. It is probable that Sebastian Cabot sailed along the shores of what was afterwards called Florida, but a few years after Columbus discovered America. Spanish authors claim that Juan Ponce de Leon discovered and named Florida, in 1512. Narvaez, another Spanish commander, having obtained a grant of Florida, in 1528, landed four or five hundred men, but was lost by shipwreck, near the mouth of the Mississippi. Ferdinand de Soto was probably the first white man who saw the Mississippi river. He is said to have marched 1000 men from Florida, through the Chickasaw country, to the Mississippi, near the mouth of Red river, where he took sick and died. His men returned. Some writers suppose De Soto travelled as far north as Kentucky, or the Ohio river. This is not probable.

The French were the first to explore and settle the West, and they held jurisdiction over the country of Illinois for eighty years, when it fell into the hands of the British upon the conquest of Canada.

In 1564, Florida was settled by a colony of Huguenots, under Admiral Coligny, who were afterwards massacred by the Spaniards, because they were Protestant *heretics!*

In 1608, Admiral Champlaine founded Quebec, from which, French settlements spread through the Canadas.

About 1670, the notion prevailed amongst the French that visited Canada, that a western passage to the Pacific ocean existed. They learned from the Indians, that far in the west there was a great river, but of its course or termination they could learn nothing. They supposed that this river communicated with the Western ocean.

To investigate this question, P. Marquette, a Jesuit, and Joliet, were appointed by M. Talon, the Intendant of New France. Marquette was well acquainted with the Canadas, and had great influence with the Indian tribes. They conducted an expedition through the lakes, up Green bay and Fox river, to the Portage, where it approaches the Wisconsin, to which they passed, and descended that river to the Mississippi, which they reached the 17h of June, 1673. They found a river much larger and deeper than it had been represented by the Indians. Their regular journal was lost on their return to Canada; but from the account afterwards given by Joliet, they found the natives friendly, and that a tradition existed amongst them of the residence of a "Mon-e-to," or spirit, near the mouth of the Missouri, which they could not pass. They turned their course up the Illinois, and were highly delighted with the placid stream, and the woodlands and prairies through which it flowed. They were hospitably received, and kindly treated, by the Illinois, a numerous nation of Indians, who were destitute of the cruelty of savages. The word "Illinois," or "Illini," is said by Hennepin, to signify a "*full grown man.*" This nation appears to have originally possessed the Illinois country, and also a portion west of the Mississippi. The nation was made up of eight tribes;—the Miamies, Michigamies, Mascotins, Kaskaskias Kahokias, Peorias, Piankeshaws, and Tau-mar-waus.

Marquette continued among these Indians, with a view to christianize them; but Joliet returned to Canada, and reported the discoveries he had made.

Several years elapsed before any one attempted to follow up the discoveries of Marquette and Joliet. M. de La Salle, a native of Normandy, but who had resided many years in Canada, was the first to extend these early discoveries. He was a man of intelligence, talents, enterprise, and perseverance. After obtaining the sanction of the king of France, he set out on his projected expedition in 1678, from Frontenac, with Chevalier Tonti, his lieutenant, and Father Hennepin, a Jesuit missionary, and thirty or forty men.

He spent about one year in exploring the country bordering on the lakes, and in selecting positions for forts and trading posts, to secure the Indian trade to the French. After he had built a fort at Niagara, and fitted out a small vessel, he sailed through the lakes to Green bay, then called the "Bay of Puants." From thence he proceeded with his men in canoes towards the south end of lake Michigan, and arrived at the mouth of the "river of the Miamis" in November, 1679. This is thought to be the Milwaukee, in Wisconsin

Territory. Here he built a fort, left eight or ten men, and passed with the rest of his company across the country, to the waters of the Illinois river, and descended that river a considerable distance, when he was stopped for want of supplies. This was occasioned by the loss of a boat, which had been sent from his post on Green bay. He was now compelled by necessity to build a fort, which, on account of the anxiety of mind he experienced, was called *Creve-cœur*, or broken heart.

The position of this fort cannot now be ascertained, but from some appearances, it is thought to have been near Spring bay, in the north-east part of Tazewell county.

At this period, the Illinois were engaged in a war with the Iroquois, a numerous, warlike, and cruel nation, with whom La Salle had traded, while on the borders of Canada. The former, according to Indian notions of friendship, expected assistance from the French; but the interests and safety of La Salle depended upon terminating this warfare, and to this object he directed his strenuous efforts. The suspicious Illinois construed this into treachery, which was strengthened by the malicious and perfidious conduct of some of his own men, and pronounced upon him the sentence of death. Immediately he formed and executed the bold and hazardous project of going alone and unarmed to the camp of the Illinois, and vindicating his conduct. He declared his innocence of the charges, and demanded the author. He urged that the war should be terminated, and that the hostile nations should live in peace.

The coolness, bravery, and eloquence of La Salle filled the Indians with astonishment, and entirely changed their purposes. The calumet was smoked, presents mutually exchanged, and a treaty of amity concluded.

The original project of discovery was now pursued. Father Hennepin started on the 28th of February, 1680, and, having passed down the Illinois, ascended the Mississippi, to the falls of St. Anthony. Here he was taken prisoner, robbed, and carried to the Indian villages, from which he made his escape, returned to Canada, by the way of the Wisconsin, and from thence to France, where he published an account of his travels.*

La Salle visited Canada to obtain supplies, returned to Creve-cœur, and shortly after descended the Illinois and then the Mississippi, where he built one or two forts on its banks, and took possession of the country in the name of the king of France, and in honor of him called it *Louisiana.*

One of these forts is thought to have been built on the west side of the river, between St. Louis and Carondalet.

After descending the Mississippi, to its

* It is difficult to determine when Hennepin writes truth or fiction. Some of his statements must be received with considerable drawback.

mouth, he returned to the Illinois, and on his way back, left some of his companions to occupy the country. This is supposed to have been the commencement of the villages of Kaskaskia and Cahokia, in 1683. La Salle went to France, fitted out an expedition to form a colony at the mouth of the Mississippi, sailed to the gulf of Mexico, but not being able to find the mouths of that river, he commenced an overland journey to his fort on the Illinois. On this journey, he was basely assassinated by two of his own men.*

After the death of La Salle, no attempts to discover the mouth of the Mississippi were made till about 1699, but the settlements in the Illinois country were gradually increased by emigrants from Canada.

In 1712, the king of France, by letters patent, gave the whole country of Louisiana to M. Crosat, with the commerce of the country, with the profits of all the mines, reserving for his own use, one fifth of the gold and silver. After expending large sums, in digging and exploring for the precious metals, without success, Crosat gave up his privilege to the king, in 1717. Soon after, the colony was granted to the Mississippi company, projected by Mr. Law, which took possession of Louis-

* La Salle appears to have discovered the bay of St. Bernard, and formed a settlement on the western side of the Colorado, in 1685.—*See J. Q. Adams's Correspondence with Don Onis. Pub. Doc., first session, fifteenth Congress*, 1818.

iana, and appointed M. Bienville, governor. In 1719, La Harpe commanded a fort, with French troops, not far from the mouth of the Missouri river.

Shortly after, several forts were built within the present limit of Illinois, of which, Fort Chartres was the most considerable. By these means, a chain of communication was formed from Canada, to the mouth of the Mississippi.

In 1699, M. Ibberville arrived in the gulf of Mexico, with two frigates, and, in March, ascended the river, in a felucca, one hundred leagues, and returned by the bayou or outlet that bears his name, through lake Ponchartrain to the gulf. He planted his colony at Biloxi, a healthy but sterile spot, between the Mobile and Mississippi rivers, and built a fortification. During several succeeding years, much exploring was done, and considerable trade carried on with the Indians, for peltries, yet these expeditions were a source of much expense to France.

In January, 1702, the colony at Mobile was planted; several other settlements were soon after formed. The Catholics also commenced several missions amongst the Indians. Difficulties frequently occurred with their Spanish neighbors in Florida and Mexico.

M. Ibberville died in 1706, and M. Bienville succeeded him in the government of Louisiana for many years. The city of New Orleans was founded, during his administration, in

1719. It is situated on the east bank of the Mississippi, one hundred and five miles from its mouth. From 1723 to 1730, the French had exterminating wars with the Natchez, a powerful nation of Indians. They had killed 700 French in 1723, and about 1730 the French exterminated the nation. Various wars took place subsequently with the Spanish and English. But over most of the Indians along the Mississippi, these French colonists gained extraordinary influence. During this period, emigrants continued to arrive from France, so that the colonists rapidly increased in numbers.

The Mississippi land scheme, or "bubble," as it was called, originated with the celebrated John Law, in 1717, which soon burst, and spread ruin throughout the moneyed interests of France. The amount of stock created, was said to equal 310,000,000 of dollars. The whole proved an entire failure, but it served to increase greatly the population of Louisiana, so that, from 1736, the colonies in the Lower Valley prospered.

In 1754, the war commenced between France and England relative to the boundaries of the Canadas. At that period, France claimed all the countries west of the Alleghany mountains, while England, on the other hand, had granted to Virginia, Connecticut and other colonies, charters which extended across the continent to the "South sea," as the Pacific ocean was then called. A grant also was made by Virginia, and the crown of Great Britain, of 600,000 acres, to a company called "The Ohio Company." The governor of New France, as Canada and Louisiana was then called, protested, erected forts on lake Erie, and at the present site of Pittsburgh, and enlisted the Indians against the English and Americans. Pittsburgh was then called Fort du Quesne. Then followed Braddock's war, as this contest is called, in the West,—the mission of Major (afterward General) Washington,—the defeat of Braddock; and, finally, by the memorable victory of Wolfe at Quebec, and the lesser ones at Niagara and Ticonderoga, and by victories of the English fleet on the ocean, the French were humbled, and, at the treaty of Paris, in 1763, surrendered all their claims to the country east of the Mississippi. Towards the close of the war, however, France, by a secret treaty, ceded all the country west of the Mississippi, and including New Orleans, to Spain, who held possession till 1803, when it was delivered to the French government under Napoleon, and by him ceded to the United States for 15,000,000 of dollars.

The English held possession of the military posts, and exercised jurisdiction over the country of Illinois, and the adjacent regions, till 1778, during the revolutionary war; when, by a secret expedition, without direct legislative sanction, but by a most enterprising, skilful, and hazardous military manœuvre, the posts of Kaskaskia, Cahokia, Fort Chartres and Vincennes were captured by Gen. GEORGE ROGERS CLARK, with a small force of volunteer Americans, and that portion of the Valley fell under the jurisdiction of Virginia.

The legislature of Virginia sanctioned the expedition of Clark, which the Executive, Patrick Henry and his council, with Thomas Jefferson, George Wythe, and George Mason, by written instructions, had agreed should be done, and a county called "Illinois" was organized the same year.

In 1784, Virginia, in conjunction with other States, ceded all claims to the Great West to the United States, reserving certain tracts for the payment of revolutionary claims. This cession laid the foundation for five new States north-west of Ohio, when each district should have 60,000 inhabitants, and even a less number, by consent of Congress. Two restrictions were peremptorily enjoined,—that each State should adopt a constitution with a republican form of government, and that slavery or involuntary servitude should be for ever prohibited.

It is unnecessary here to enter into details of the settlement of each particular State,—the incessant attacks from the Indians,—the border wars that ensued,—the adventures of Boone and his associates in settling Kentucky,—the unfortunate campaigns of Harmar and St. Clair,—the victorious one of Wayne,—or the reminiscences and events of the war of 1812, and its termination in 1815. Some historical notices of each State may be found in their proper place.

Prospective Increase of Population. For a long period, in the States of the West, the increase of population was slow, and retarded by several causes. Difficulties of a formidable character had to be surmounted. The footsteps of the American emigrants were every where drenched in blood, shed by infuriated savage foes, and before 1790, more than 5000 persons had been murdered, or taken captive and lost to the settlements. "It has been estimated, that, in the short space of *seven* years, from 1783 to 1790, more than fifteen hundred of the inhabitants of Kentucky were either massacred or carried away into a captivity worse than death, by the Indians; and an equal number from Western Pennsylvania and Virginia, in the same period, met with a similar fate. The settlers on the frontiers were almost constantly, for a period of forty years, harassed either by actual attacks of the savages, or the daily expectation of them. The tomahawk and the scalping-knife were the objects of their fears by day and by night."*

Hence, in suggesting reasons showing why the population of this Valley must increase in future in a far greater ratio than in the past, it will appear,

1. That the most perfect security is now

* Baird.

enjoyed by all emigrants, both for their families and property.

By the wise and beneficent arrangement of government, the Indian tribes have nearly all removed to the Territory specially allotted for their occupancy west of Missouri and Arkansas. The grand error committed in past times in relation to the Indians, and which has been the source of incalculable evils to both races, has been the want of definite, fixed and permanent lines of demarcation betwixt them. It will be seen, under the proper head, that a system of measures is now in operation that will not only preserve peace between the frontier settlements and the Indian tribes, but that, to a great extent, they are becoming initiated into the habits of civilized life. There is now no more danger to the population of these States and Territories from *Indian* depredations, than to the people of the Atlantic States.

2. The increased facilities of emigration, and the advantage of sure and certain markets for every species of production, furnish a second reason why population will increase in the Western Valley beyond any former period.

Before the purchase of Louisiana, the western people had no outlet for their produce, and the chief mode of obtaining every description of merchandise,—even salt and iron,—was by the slow and expensive method of transportation by wagons and pack-horses, across almost impassable mountains and extremely difficult roads. Now, every convenience and luxury of life is carried, with comparative ease, to every town and settlement throughout the Valley, and every species of produce is sent off, in various directions, to every port on earth, if necessary. And these facilities are multiplying and increasing every hour. Turnpike roads, rail-roads, canals, and steam-boat navigation have already provided such facilities for removing from the Atlantic to the Western States, that no family desirous of removing, need hesitate or make a single inquiry as to facilities of getting to this country.

3. The facilities of trade and intercourse between the different sections of the Valley, are now superior to most countries, and are increasing every year. And no country on earth admits of such indefinite improvement, either by land or water. More than twenty thousand miles of actual steam-boat navigation, with several hundred miles of canal navigation, constructed or commenced, attest the truth of this statement. The first steam-boat on the western waters was built at Pittsburgh in 1811, and not more than seven or eight had been built, when the writer emigrated to this country in 1817. At this period (January, 1836), there are several hundred boats on the western waters, and some of the largest size. In 1817, about twenty barges,

averaging about one hundred tons each, performed the whole commercial business of transporting merchandise from New Orleans to Louisville and Cincinnati. Each performed one trip, going and returning within the year. About 150 keel-boats performed the business on the Upper Ohio to Pittsburgh. These averaged about 30 tons each, and were employed one month in making the voyage from Louisville to Pittsburgh. Three days, or three days and a half, is now the usual time occupied by the steam-packets between the two places, and from seven to twelve days between Louisville and New Orleans. Four days is the time of passing from the former place to St. Louis.

4. A fourth reason why population will increase in future in a greater ratio than the past, is derived from the increase of population in the Atlantic States, and the greater desire for removal to the West. At the close of the revolutionary war, the population of the whole Union but little exceeded two millions. Vast tracts of wilderness then existed in the old States, which have since been subdued, and from whence thousands of enterprising citizens are pressing their way into the Great Valley. Two thirds of the territory of New York, large portions of New Hampshire, Vermont and Maine, an extensive district in Middle Pennsylvania, to say nothing of wide regions in the Southern States, were comprised in this wilderness. These

extensive regions have become populous, and are sending out vast numbers of emigrants to the West. Europe is in commotion, and the emigration to North America, in 1832, reached 200,000, a due proportion of which settle in the Western Valley.

5. A fifth reason will be founded upon the immense amount of land for the occupancy of an indefinite number of emigrants, much of which will not cost the purchaser over *one dollar and twenty-five cents per acre.* Without giving the extravagant estimates that have been made by many writers, of the wide and uninhabitable desert between the Indian Territory west of Missouri and Arkansas and the Rocky mountains, nor swampy and frozen regions at the heads of the Mississippi river and around lake Superior, I will merely exhibit the amount of lands admitting of *immediate* settlement and cultivation, within the boundaries of the new States and organized Territories.

According to the report of the Secretary of the Treasury up to the 30th day of September, 1831, the estimated amount of unsold lands, on which the foreign and Indian titles had been extinguished, within the limits of the new States and Territories, was 227,293,884 acres;—and that the Indian title remained on 113,577,869 acres* within the same

* See Mr. Clay's Report on the Public Lands, April 26, 1832, U. S. Papers.

limits. The Commissioner of the General Land Office, in December, 1827, estimated the public domain, beyond the boundaries of the new States and Territories, to be 750,000,000 of acres. Much of this, however, is uninhabitable.

According to the Report of 1831, there had been granted to Ohio, Indiana, Illinois, and Alabama, for internal improvements, 2,187,665 acres;—for colleges, academies, and universities, in the new States and Territories, 508,009;—for education, being the thirty-sixth part of the public lands appropriated to common schools, 7,952,538 acres; —and for seats of government to some of the new States and Territories, 21,589 acres. Up to January, 1826, there had been sold, from the commencement of the land system, only 19,239,412 acres. Since that period, to the close of 1835, there have been sold, about 33,000,000 of acres, making in all sold, a little more than 52,000,000. This statement includes Alabama and Florida, which we have not considered as strictly within the Valley. After a hasty and somewhat imperfect estimate of the public lands, that are now in market, or will be brought into market within a few years, within the limits of Ohio, Indiana, Illinois, Missouri, Mississippi, Louisiana, Arkansas, Michigan, and the Territory of Wisconsin, the amount may be put at 130,000,000 of acres. This amount admits of immediate settlement and cultivation, and much of it may be put under cultivation without the immense labor of clearing and subduing forest lands.

The comparison between the amount of sales of public lands within the last ten years, and the preceding forty years, shows that emigration to the West is increasing at a ratio beyond what is ordinarily supposed, and that the next ten years will find a majority of the population of the United States within this Great Valley.

Sales of Land,

From 1786 to 1826,	(forty years)	19,239,412	acres.
" 1826 " 1835,	(ten years)	33,000,000	"

Three millions of families may find farms in the West.

The extensive prairie lands of Illinois and Missouri present no obstacle to the settlement of the country. Already, prairies, for many miles in extent, have been turned into farms.

6. A sixth reason why the increase of the future population of the Valley will greatly exceed the past, is derived from the increased confidence of the community in the general health of the country. The most unreasonable notions have prevailed abroad relative to the health of the Western States. All new settlements are more or less unfavorable to health, which, when cultivated and settled, become healthy. As a separate chapter will be devoted to this subject, I only advert to the fact now, of the increased confidence of the people in the Atlantic States, in the salu-

brity of our western climate, which, already, has tended to increase emigration; but which, from facts becoming more generally known, will operate to a much greater extent in future.

7. I will only add, that there is already a great amount of intelligence, and of excellent society, in all the settled portions of the Western Valley.

"The idea is no longer entertained by eastern people, that going to the West, or the 'backwoods,' as it was formerly called, is to remove to a heathen land, to a land of ignorance and barbarism, where the people do nothing but rob, and fight, and gouge! Some parts of the West have obtained this character, but most undeservedly, from the *Fearons*, the [Basil] *Halls*, the *Trollopes*, and other ignorant and insolent travelers from England, who, because they were not allowed to insult and outrage as they pleased, with Parthian spirit, hurled back upon us their poisoned javelins and darts, as they left us. There is, indeed, much destitution of moral influence and means of instruction in many, very many, neighborhoods of the West. But there is, in all the principal towns, a state of society, with which the most refined, I was going to say the most fastidious, of the eastern cities, need not be ashamed to mingle."*

The eastern emigrant will find, that wholesome legislation, and much of the influence

* Baird.

of religion, are enjoyed in the Valley of the Mississippi; extending to him all he can ask in the enjoyment of his rights, and the protection of his property.

Common school systems have been commenced in some of the States; others are following their example, and the subject of general education is receiving increasing attention every year. Colleges, and other literary institutions, are planted; and religious institutions, and means of religious instruction, are rapidly increasing. Noble and successful efforts are making by the Bible, missionary, tract, Sabbath school, temperance, and other societies, in the West. Great and rapid changes are taking place, if not to the extent we desire, yet corresponding in a degree with the gigantic march of emigration and population. Many other reasons might be urged to show that its prospective increase of population will vastly exceed the ratio of its retrospective increase, but these are sufficient.

CHAPTER III.

CLIMATE.

Comparative View of the Climate with the Atlantic States —Diseases—Means of preserving Health.

In a country of such vast extent, through fifteen degrees of latitude, the climate must necessarily be various. Louisiana, Mississippi, and the lower half of Arkansas, lie between the latitudes of 30° and 35°, and correspond with Georgia and South Carolina. Their difference of climate is not material. The northern half of Arkansas, Tennessee, and Kentucky, lie west from North Carolina and the southern portion of Virginia. The climate varies from those States only as they are less elevated than the mountainous parts of Virginia and Carolina. Hence, the emigrant from the southern Atlantic States, unless he comes from a mountainous region, will experience no great change of climate, by emigrating to the Lower Mississippi Valley. Missouri, Illinois, Indiana, and Ohio, lie parallel with the northern half of Virginia, Maryland, Delaware, Pennsylvania, New Jersey, and so much of New York and New England as lies south of latitude 42° north. But several circumstances combine to produce variations in the climate:

1. Much of those Atlantic States are hilly, and, in many parts, mountainous, some of which are two and three thousand feet above the level of the ocean. The parallel Western States have no mountains, and are not proportionably hilly.

2. The Atlantic States border on the ocean, on the east, and feel the influence of the cold, damp winds, from the north-east and east. Their rains are more copious, and their snows deeper. The northern portions of the West, equally with New York and Vermont, are affected with the influence of the lakes, though not to the same extent.

3. "The courses of rivers, by changing, in some degree, the direction of the winds, exert an influence on the climate. In the Atlantic States, from New England to North Carolina, the rivers run more or less to the south-east, and increase the winds which blow from the north-west, while the great bed of the Mississippi exerts an equal influence in augmenting the number and steadiness of the winds which blow over it from the south-west; and there is another cause of difference in climate, chiefly perceptible, first, in the temperature, which, if no counteracting cause

existed, they would raise in the West, considerably above that of corresponding latitudes in the East; and, secondly, in the moisture of the two regions, which is generally greater west than east of the mountains, when the south-west wind prevails; as much of the water, with which it comes charged from the gulf of Mexico, is deposited, before it reaches the country east of the Alleghanies."*

It is an error, that our climate is more variable, or the summers materially hotter, than in a corresponding latitude in the Atlantic States. "The New Englander and New Yorker, north of the mountains of West Point, should bear in mind, that his migration is not to the *west*, but *south-west;* and as necessarily brings him into a warmer climate, as when he seeks the shores of the Delaware, Potomac, or James river."

The settlers from Virginia to Kentucky, or those from Maryland and Pennsylvania to Ohio, or farther west, have never complained of hotter summers than they had found in the land from whence they came.

To institute a comparative estimate of temperature, between the East and the West, we must observe,—first, the thermometer; and, secondly, the flowering of trees, the putting forth of vegetation, and the ripening of fruits and grain, in *correspondent latitudes.* This has not usually been done. Philadelphia and

* Dr. Drake.

Cincinnati approach nearer to the same parallel, than any other places where such observations have been made. Cincinnati, however, is about 50′ south of Philadelphia. The following remarks are from Dr. Daniel Drake, of Cincinnati, to whose pen the West is much indebted:

"From a series of daily observations in Cincinnati, or its vicinity, for eight consecutive years, the mean annual temperature has been ascertained to be 54 degrees and a quarter. Dr. Rush states the mean temperature of Philadelphia at 52 degrees and a half; Dr. Coxe, from six years' observations, at 54 degrees and a sixth; and Mr. Legaux, from seventeen years' observations, at Spring Mill, a few miles out of the city, at 53 degrees and a third; the mean term of which results, 53 degrees and a third, is but the fraction of a degree lower than the mean heat of Cincinnati, and actually less than should be afforded by the difference of latitude.

"A reference to the temperatures of summer and winter, will give nearly the same results. From nine years' observations (three at Spring Mill, by Mr. Legaux, and six in Philadelphia, by Dr. Coxe), the mean summer heat of that part of Pennsylvania appears to be 76 degrees and six tenths. The mean summer heat at Cincinnati, for an equal number of years, was 74 degrees and four tenths. The average number of days in which the thermometer rose to 90 degrees or upwards, during the same period, was fourteen each summer; and the greatest elevation observed, was 98 degrees; all of which would bear an almost exact comparison with similar observations in Pennsylvania. Mr. Legaux states the most intense cold, at Spring Mill, from 1787 to 1806, to have been 17 degrees and five tenths below cipher; while, within the same period, it was 18 degrees at Cincinnati. The average of extreme cold, for several years, as observed by Mr. Legaux, was one degree and eight tenths below cipher: the same average at Cincinnati, was two degrees below. From all which, we may conclude that the banks of the Delaware and Ohio, in the same latitudes, have nearly the same temperature."

The State of Illinois, extending, as it does, through five and a half degrees of latitude, has considerable variation in its climate. It has no mountains, and, though undulating, it cannot be called hilly. Its extensive prairies and level surface, give greater scope to the winds, especially in winter. In the southern part of the State, during the three winter months, snow frequently falls, but seldom lies long. In the northern part, the winters are as cold, but not so much snow falls, as in the same latitudes in the Atlantic States.

The Mississippi, at St. Louis, is frequently frozen over, and is crossed on the ice, and, occasionally, for several weeks. The hot season is longer, though not more intense, than occasionally, for a day or two, in New England.

During the years 1817, '18, '19, the Rev. Mr. Giddings, at St. Louis, made a series of observations upon Fahrenheit's thermometer:

	Deg.	Hund.
Mean temperature for 1817,	55	52
Mean temperature from the beginning of May, 1818, to the end of April, 1819,	56	98
Mean temperature for 1820,	56	18

The mean of these results is about fifty-six degrees and a quarter.

The mean temperature of each month, during the above years, is as follows:

	Deg.	Hund.
January,	30	62
February,	38	65
March,	43	13
April,	58	47
May,	62	66
June,	74	47
July,	78	66
August,	72	88
September,	70	10
October,	59	00
November,	53	13
December,	34	33

The mean temperature of the different seasons is as follows:

	Deg.	Hund.
Winter,	34	53
Spring,	54	74
Summer,	74	34
Autumn,	60	77

The greatest extremes of heat and cold during my residence of eighteen years, in the vicinity of St. Louis, is as follows:

Greatest heat in July, 1820, and July, 1833, 100 degrees. Greatest cold, Jan. 3d, 1834, 18 degrees below zero; Feb. 8th, 1835, 22 degrees below zero.

The foregoing facts will doubtless apply to about one half of Illinois. This climate, also, is subject to sudden changes, from heat to cold, from wet to dry, especially from November to May. The heat of the summer, below the fortieth degree of latitude, is more enervating, and the system becomes more easily debilitated, than in the bracing atmosphere of a more northerly region.

At Marietta, Ohio, in latitude 39° 25′ north, and at the junction of the Muskingum river with the Ohio, the mean temperature, for 1834, was 52 degrees and four tenths; highest, in August, 95 degrees; lowest, January, at zero. Fair days, 225; cloudy days, 110.

At Nashville, Tenn., 1834, the mean temperature was 59 degrees and seventy-six hundredths; maximum 97, minimum four above zero. The summer temperature of this place never reaches 100 degrees. January 26th, 1832, 18 degrees below zero; February 8th, 1835, 10 degrees below zero.

The putting forth of vegetation in the spring, furnishes some evidence of the character of the climate of any country, though by no means entirely accurate. Other causes combine to advance or retard vegetation. A wet or dry season, or a few days of heat or cold, at a particular crisis, will produce material changes.

The following table is constructed from memoranda, made at the various dates given, near the latitude of St. Louis, which is computed at 38° 30′. The observations of 1819, were made at St. Charles, and vicinity, in the State of Missouri. Those of 1820, in St. Louis county, 17 miles north-west from the city of St. Louis. The remainder, at Rock Spring, Illinois, 18 miles east from St. Louis. It will be perceived, the years are not consecutive. In 1826, the writer was absent to the Eastern States, and, for 1828, his notes were too imperfect to answer the purpose.

In the column showing the times of the first snows, and the first and last frosts in the season, a little explanation may be necessary. A "light" snow means merely enough to whiten the earth, and which usually disappears in a few hours.

Many of the frosts recorded "light," were not severe enough to kill ordinary vegetation.

Year.	Peach and Red bud in blossom.	Strawberries in blossom.	Blackberries in blossom.	Apple leaves begin to put forth.	Apple trees in blossom.	Grass green in prairies.	Oaks and other forest trees put forth leaves.	First snow on approach of winter.	Last frost in Spring.	First frost in Autumn.
1819	April 4.	not noted	May 19.	April 15.	April 20.	April 18.	Half size, May 19.	Oct. 8, few flakes.	May 18, very light.	Sept. 23.
1820	Ap.14.no peach B.	April 2.	May 10. fall off, 17	Mar. 25.	April 15.	April 10.	April 22. full size, May 7.	Oct.24.few fl'ks Nov. 11, 3 inch.	June 1, very light.	Sept. 20. Oct.8, ice.
1821	Ap.26.no peach B.	April 30.	May 21.	April 24.	May 3.	April 26.	Ap.26 to May 3 f.grown, 22.	Nov. 8, 2½ inches.	Ap.18, severe. May 9, light.	Oct. 8.
1822	April 5.	April 25.	May 10.	April 18.	April 22.	April 10.	Ap.29. f.size, May 14.	Nov. 16, light.	April 16, severe, ice.	Oct. 13.
1823	April 19.	April 26.	May 20.	April 15.	April 28.	April 10.	April 23.	Nov. 1, light.	April 24.	Sept. 21–2. Ice 23.
1824	April 20.	April 28.	May 18.	April 20.	April 29.	April 14.	April 30.	Nov. 7.	May 5.	Oct. 21. hard frost.
1825	Mar. 25.	Ap 3. ripe May 17.	May 8.	Mar. 30.	April 5.	Mar. 10.	April 3.	Dec. 11, 3 inches.	Feb.22. Next. April 20, ice.	Oct. 2–3. 27th, ice.
1827	April 4.	April 10.	May 15.	April 4.	April 13.	Mar. 25.	April 10. full size, April 30.	Nov. 25, light.	May 7, light.	Sept. 23, light.
1829	April 20.	April 24.	May 20.	April 20.	April 26.	April 24.	April 27.	Nov. 12, four inches sleet.	not noted.	Sept. 17.
1830	April 1.	April 5.	May 9.	April 1.	April 9.	April 1.	begin April 5. f. size, May 1.			

These observations, upon a comparison with the same parallels of latitude in the Eastern States, show that there is no material difference of climate between the two sections of our country, except that produced by local causes, as mountainous districts, contiguity to the ocean, &c.

A similar error has existed in relation to sudden and extreme changes of weather in the West. People who emigrate to a new country, have their curiosity awakened, and, perhaps, for the first time in their lives, become quite observing of such changes. From habitually observing the weather, the impression is produced on their minds that there is a marked difference in this climate. Dr. Rush declares that there is but *one* steady trait in the character of the climate of Pennsylvania, and that is, *it is uniformly variable;* and he asserts that he has known the thermometer to fall 20 degrees in one hour and a half. March 26–27, 1818, the thermometer, in St. Louis, fell 41 degrees in thirty hours,—from 83 degrees to 42 degrees. I have no record or recollection of a more sudden change, in eighteen years. Mr. Legaux saw it fall, in the vicinity of Philadelphia, 47 degrees in twenty-four hours; and Dr. Drake states that this is five degrees more than any impression ever observed in Cincinnati, in the same length of time. Emigrants from New England and the northern part of New York

State must not expect to find the same climate in the West, at 38 or 40 degrees; but let them remove to the same parallel of latitude in the West, to Wisconsin, or the northern part of Illinois, and they will probably find a climate far more uniform than the land of their birth.

Prevailing winds modify and affect the climate of every country. South-westwardly winds prevail along the Mississippi Valley. The following tabular view of observations made at Cincinnati, by Dr. D. Drake, for six succeeding years, with so few omissions, that they amount to 4200, will give further illustrations of this subject. They have been brought from eight points of the compass:

Observations.

MONTHS.	S. E.	S.	S. W.	N. E.	N.	N. W.	E.	W.	CALM.
January, ...	6	2	13	8	1	21	3	6	6
February, ...	5	1	13	8	1	14	0	5	8
March, ...	10	1	16	11	1	10	0	5	4
April, ...	7	1	24	10	1	8	1	3	5
May, ...	7	1	19	10	0	10	1	4	6
June, ...	9	1	23	12	5	7	1	2	3
July, ...	6	1	19	11	2	11	1	4	4
August, ...	6	1	23	10	1	12	1	1	6
September, ...	6	1	23	9	0	8	2	3	3
October, ...	9	1	24	6	1	10	2	4	3
November, ...	9	3	13	6	1	10	2	7	5
December, ...	7	1	11	5	0	15	2	6	9
Total,	87	14	221	106	14	136	16	50	62

The results of my own observations, made for twelve years, with the exception of 1826, and with some irregularity, from traveling in different parts of Missouri and Illinois, during the time, do not vary in any material degree from the above table, excepting fewer east and north-east winds.

Dr. Drake has given a table, setting forth the results of 4268 observations on the state of the weather at Cincinnati, from which it will be perceived that of the 365 days in a year, about 176 will be fair, 105 cloudy, and 84 variable.

Dr. L. C. Beck made similar observations at St. Louis, during the year 1820, which produced the result of 245 clear days, and cloudy, including variable days, 110.

Years.	Clear days.	Cloudy days.	Variable days.
1	180	107	68
2	158	112	91
3	187	78	85
4	152	106	107
5	185	111	68
6	172	112	74
Total, 6	1034	626	493
Mean terms,	172. 33	104. 33	82. 16

The following table shows the condition of the weather, in each month of a mean year, for the above period:

Months.	Clear days.	Cloudy days.	Variable days.
January,	9. 8	13. 1	7. 8
February,	10. 3	12. 0	6. 5
March,	13. 5	9. 1	8. 3
April,	13. 1	10. 8	7. 6
May,	15. 0	8. 5	7. 5
June,	15. 5	5. 0	9. 6
July,	19. 0	5. 5	6. 0
August,	19. 6	4. 6	6. 5
September,	19. 5	5. 3	6. 1
October,	16. 1	6. 0	8. 1
November,	9. 5	13. 5	5. 5
December,	9. 6	14. 1	5. 8

There would be some variations from the foregoing table, in a series of observations in the country bordering upon the Upper Mississippi and Missouri. The weather in the States of Ohio and Kentucky, is, doubtless, more or less affected in autumn, by the rains that fall on the Alleghany mountains, and the rise of the Ohio and its tributaries. So the weather in the months of April, May and June, in Missouri, is affected by the spring floods of the Missouri and Mississippi rivers.

The following table is constructed from a series of observations made at the military posts in the West, by the surgeons of the United States army, for four years,—1822, 1823, 1824 and 1825. (See *American Almanac* for 1834, p. 81.)

Posts.	Situations.	North latitude ° ′	Elevation above ocean.	Mean Temp. for four years.	Maximum.	Minimum.	Range of Thermometer.	Weather. Monthly average. Fair days.	C'l'dy days.	Rain. days.	Sn'w days.
Fort Brady,	Sault de St. Mary, outlet of Lake Superior, ...	46 22	5 95	41 37	90	—33	1 23	13 30	2 27	7 83	6 02
Fort Snelling,	Mouth of St. Peters, 10 m. below St. Anthony falls,	46 39	7 80	45 00	96	—29	1 25	16 94	5 50	5 77	2 22
Fort Howard,	Green Bay, Wisconsin T.	45 00	6 00	44 50	1 00	—38	1 38	15 47	7 98	4 56	2 42
FortCrawford,	Prairie du Chein, Wisconsin Territory, ...	43 25	5 80	45 52	96	—28	1 24	16 80	6 29	3 87	1 32
Council Bluffs,	Upper Missouri, ...	41 31	8 00	50 82	1 08	—21	1 29	19 68	6 54	2 95	1 25
Cantonment Jessup,	On Red River, La., ...			68 31	97	7	90	18 63	4 49	7 25	05
Baton Rouge,	Louisiana, ...	30 32		68 07	99	18	81	20 16	4 08	6 16	

— *signifies below zero.*

The times of observation at the above posts, were 7 A. M. and 2 and 9, P. M. The mean of each month was deduced from ninety observations, and of each year from one thousand and ninety-five observations. The reader, who is desirous of following up this comparative view of the climate between the Atlantic States and the Valley of the Mississippi, can compare the observations recorded in these tables, with similar observations made in the same parallels of latitude. He will find the climate of the West quite as uniform, and the weather as little variable, as in the Atlantic States.

Diseases, means of preserving Health, &c. Of the Lower Valley, I shall say but very little. Dr. Drake observes: "The diseases of this portion of the great Valley are few, and prevail chiefly in summer and autumn. They are the offspring of the combined action of intense heat and marsh exhalation." They are, generally, remittent and intermittent bilious fevers. Emigrants most generally undergo a seasoning, or become acclimated. Many persons, however, from the Northern and Middle States and from Europe, enjoy health. In sickly situations, these fevers are apt to return, and often prove fatal. They frequently enfeeble the constitution, and produce chronic inflammation of the liver, enlargement of the spleen, or terminate in jaundice or dropsy, and disorder the digestive organs. When persons find themselves subject to repeated attacks, the only safe resource is an annual migration to a more northern climate, during the summer. Many families from New Orleans, and other exposed situations, retire to the pine barrens of Louisiana, in the hot and sickly season, where limpid streams, flowing over a pebbly bed, and a terebinthine atmosphere, are enjoyed. Eight months of the year are pleasant and healthy in the Lower Mississippi Valley.

The advice of Dr. Drake is, that "Those who migrate from a colder climate to the southern Mississippi States, should observe the following directions: 1st. To arrive there in autumn, instead of spring or summer. 2d. If practicable, to spend the hottest part of the first two or three years in a higher latitude. 3d. To select the healthiest situations. 4th. To live temperately. 5th. To preserve a regular habit. Lastly, to avoid the heat of the sun, from ten in the morning till four in the afternoon; and, above all, the night air. By a strict attention to these rules, many would escape the diseases of the climate, who annually sink under its baleful influence."

Those States and Territories to which this work is intended more immediately as a Guide, do not differ very materially in salubrity. The same general features are found in each. There is but little diversity in climate; their geological and physical structure coincide, and the experience of many years shows that there is no great difference. Where autumnal fevers are common, they are usually of similar character. The same causes for disease, exist in Ohio as in Missouri; in Michigan as in Illinois; in Kentucky and Tennessee as in Indiana. All these States are much more infested with the maladies which depend on variations of temperature, than the States farther south. All have localities where intermittents and agues are found, and all possess extensive districts of country where health is enjoyed by a very large proportion of emigrants. There is some difference between a heavily timbered and a prairie country, in favor of the latter, other circumstances being equal. Changes, favorable to continued health, are produced by the settlement and cultivation of any particular portion of country. Of one fact, I have long since satisfied my mind,—that ordinary fevers are not caused by the use of the water of the West.

Exceptions may be made in some few cases, where a vein of water is impregnated with some deleterious mineral substance. The use of a well, dug in the vicinity of a coal-bed in Illinois, was supposed to have caused sickness in a family, for two seasons. Any offensive property in water is readily detected by the taste. Cool, refreshing water is a great preservative of health. It is common for families, who aro too indifferent to their comfort to dig a well, to use the tepid, muddy water of the small streams, in the frontier States, during the summer, or to dig a shallow well, and wall it with timber, which soon imparts an offensive taste to the water. Water of excellent quality, may be found in springs, or by digging from twenty to thirty feet, throughout the Western States. Most of the water, thus obtained, is hard water, from its limestone qualities, but it is most unquestionably healthy. Those persons who emigrate from a region of sandstone, or primitive rock, where water is soft, will find our limestone water to produce a slight affection of the bowels, which will prove more advantageous to health, than otherwise, and which will last but a few weeks. Whenever disease prevails in the Western States, it may, generally, be attributed to one or more of the following causes:

1. *Variations of temperature.* This cause, we have already shown, exists to as great extent in the same latitude east of the mountains.

2. *The rapid decomposition of vegetable matter.* In all our rich lands, there are vast quantities of vegetable matter mixed with the soil, or spread over the surface. Extreme hot weather, following especially a season of much rain, before the middle of July, will produce sickness. If the early part of summer be tolerably dry, although a hot season follows, sickness does not generally prevail. The year 1820 was an exception to this rule. It was a very dry, hot, sickly year through the West, indeed, throughout the world. A wet season, with a moderately cool atmosphere, has proved healthy.

3. *Marsh exhalations.* These, combined with heat, will always generate fevers. Indeed, there is probably very little difference in the miasm thrown off from decomposed vegetable matter, and that produced from sluggish streams, standing waters and marshes. These, in the great Valley, abound with decayed vegetable matter. Hence, along the streams which have alluvial bottoms (as low lands, upon streams, are called in the West), some of which are annually overflowed, and where the timber and luxuriant vegetable

growth are but partially subdued, the inhabitants are liable to fevers, dysenteries and agues. Situations directly under the bluffs, adjacent to the bottom lands, that lie upon our large rivers, especially when the vegetation is unsubdued, have proved unhealthy. So have situations at the heads or in the slope of the ravines that put down from the bluffs towards the rivers.

The principal diseases that prevail, may be stated as follows: In the winter, and early in the spring, severe colds, inflammation of the lungs and pleurisies are most common. The genuine hereditary consumption of New England is rare, and families and individuals predisposed to that disease, might often be preserved, by migration to this Valley. Acute inflammation of the brain, and inflammatory rheumatism are not unusual at that season.

During the summer and autumn, cholera infantum with children in large towns, diarrhœa, cholera morbus, dysentery, intermittent and remittent bilious fevers prevail. The intermittent assumes various forms, and has acquired several names amongst the country people, where it prevails more generally than in large towns. It is called the "chill and fever,"—"ague,"—"dumb ague," &c., according to its form of attack.

The remittent fever is the most formidable of our autumnal diseases, especially when of a highly bilious type. In most seasons, these diseases are easily managed, and yield to a dose or two of medicine. Sore eyes, especially in autumn, is a common complaint in the frontier settlements, and when neglected or improperly managed, have terminated in total blindness.

The "milk sickness," as it is called, occasionally prevails in some localities, some particulars of which will be found in another place. There is a disease that afflicts many frontier people, called by some "sick stomach," by others, "water brash," from its symptoms of sudden nausea, with vomiting, especially after meals.

In 1832, the cholera made its appearance in the West. In many places, its first approach was attended with great mortality, but its second visit to a place has been in a milder and more manageable form. It has visited various parts of the West, on each returning season since, especially along the great rivers and about the steam-boats.* It appears to have changed somewhat the characteristics of our western diseases, and will probably become a modified and manageable disease. Since its visit, our fevers are more congestive, less bile is secreted, and the stomach more affected. The subject will doubtless be noticed by our physicians, and observations made, how far this new disease will become assimilated to the ordinary diseases of the country.

* Not a case of cholera has occurred this season,—Aug. 12, 1836.

We are satisfied, after a long course of observations, much traveling, and conversing with many hundreds of families, with the view of arriving at correct conclusions on these subjects, *that there is no such operation as that of emigrants undergoing a seasoning, or becoming acclimated*, in the States of Ohio, Indiana, Illinois, Missouri, Kentucky, Tennessee, Michigan, or the Wisconsin Territory. *Nor does it make the least difference from what part of the United States or Europe, they come, nor whether they arrive here in the spring or autumn.* There is an erroneous notion prevailing in some of the Atlantic States, on this subject, that should be corrected. When sickness prevails, there is just as much, and it is equally severe, amongst the old settlers, those born in the country, or who migrate from the Carolinas or Georgia, as those who come from the Northern States. Families are just as liable to sickness, and are as often attacked for the first time, after residing several years in the country, as at any other time. A large proportion of the families and individuals, who remove from New England, to the various parts of the Valley, north of the 37th degree of latitude, have no sickness the first year.

The impression has formerly existed abroad, that Illinois is less healthy than other Western States. This is entirely erroneous. As in all countries, there are some localities, where the causes that produce sickness exist more than in others. This is not the fact with Illinois in general.

That this State is as healthy as any other Western State, can be abundantly supported by facts. Let a candid observer compare the health of the early settlers of New England, with that of the early settlers of the West, and he will find the scale to preponderate in favor of the latter. Unless there is some strange fatality attending Illinois, its population must be more healthy than the early settlers of a timbered region. But in no period of its history have sickness and death triumphed, in any respect, equal to what they did two or three years since, in the lake country of New York.

The year 1811, is recorded in the memoirs of the early settlers, as a season of unusual sickness, near the banks of the Mississippi and Missouri rivers. The latter rose to an unusual height, in June; the waters of the small creeks were backed up, and a large surface of luxuriant vegetation was covered and deadened. This was succeeded by hot and dry weather. Bilious and intermittent fevers prevailed extensively. The seasons of 1819, '20, and '21, were unusually sickly in Illinois and Missouri. Emigrants, in shoals, had spread over a wide range of country, within a year or two preceding. Multitudes were placed under circumstances the most unfavorable to the preservation of health, in new and open cabins of green timber, often using the stagnant water of creeks and ponds, with a luxuriant vegetation around them undergoing decomposition, and all the other evils attendant on the settlement of a new and unbroken country. Under such circumstances, can it be surprising that many were sick, and that many died? The summer of 1820 was the hottest and driest ever known in this country. For weeks, in succession, the thermometer, in the shade, at St. Louis, was up to 96 degrees for hours in the day. Not a cloud came over the sun, to afford a partial relief from its burning influence. The fevers of that season were unusually rapid, malignant and umanageable. Almost every mark of the yellow fever, as laid down in the books, was exhibited in many cases, both in town and country. The bilious fever put on its most malignant type. Black, fœtid matter was discharged from the stomach and by stools. The writer and all his family suffered severely that season. He lived seventeen miles from St. Louis, on the road to St. Charles, in Missouri, on a farm; the settlement had been called healthy. The Missouri bottom was one mile distant. Three miles west-south-west, was the Creve-cœur lake, a body of water several miles in length and half a mile in width, connected by an outlet, with the Missouri river. The water of this lake was entirely stagnant, covered with a thick scum, and sent forth a noisome smell: fish in it died. My oldest son, a robust youth, ten years of age, and my brother-in-law, a hale and stout young man, sickened and died the first week in October. I was attacked the fifth of July, and came as near dying as a person could and recover. All my children were sick. While convalescent, in September, I took a long journey to Cape Girardeau country, one hundred and twenty miles south, and back through the lead-mine country, to the Missouri river, sixty miles west of St. Louis, and in all the route found that sickness had prevailed to the same extent.

At Vincennes, and other parts of Indiana, disease triumphed. The country around Vincennes, on the east side of the Wabash, is a sandy plain. A gentleman who escaped the ravages of fever in that place, and who was much engaged in nursing the sick and consoling the dying, stated to me that nothing was so disheartening as the cloudless sky and burning sun, that continued unchanged, for weeks in succession. Mortality prevailed to a great extent along the banks of the Wabash. Hindostan, a town on the east fork of White river, thirty-eight miles from Vincennes, on the road to Louisville, was begun the preceding year. Seventy or eighty families had crowded in, at the commencement of the year 1820. The heavy timber of poplar (whitewood), oak and beech, had been cut down, the brush burned, and the logs left on the ground. By June, the bark was loosened, an intolerable stench proceeded from the timber; sickness followed, and about two thirds of the population died. And yet, to look about the place, there is no local cause that would indicate sickness. In the summer of 1821, sickness prevailed very extensively, but in a much milder form: its type was intermittent, and usually yielded to ordinary remedies. During that year, the number of deaths in St. Louis, was 136,—the population 5000. At least one third of that number were strangers and transient persons, who either arrived sick, or were taken sick within two or three days after arrival. St. Louis had then no *police* regulations, the streets were filthy in the extreme, and the population were crowded into every hole and corner. This was the most sickly and dying season St. Louis ever knew, except when the cholera prevailed, in October, 1832.

The same years (1820–21) were noted for unusual sickness throughout the United States, and indeed the whole world. The bilious fever prevailed in the hilly and mountainous districts of Virginia and Pennsylvania, and even among the Green mountains of Vermont.

Very little general sickness (except cholera, in 1832–33) prevailed in 1830, '31, '32, or '33. In 1834, congestive fever, and dysentery, with some of the symptoms of cholera, existed in many places in the West, though not extensively fatal. In the month of June, were frequent sudden showers in Illinois and Missouri, with intervals of extreme heat; July and August very hot and dry. The disease began early in July, and continued till September.

The year 1835, was the most sickly year,

for common intermittents, *which prevailed more amongst the old settlers, than the newly arrived emigrants.* In Illinois, and generally throughout the West, below the 40th degree of latitude, it was sickly, though not fatal. Early in the spring, till the month of May, it was unusually dry, and vegetation was two weeks later than usual. May, and a part of June, were very wet, followed by a few days of extremely hot weather. Vegetation grew with great luxuriance. Newly ploughed ground sent forth a noxious effluvium, with a most offensive odor, and after a few days would be covered with a greenish coat, like the scum on stagnant water. Town situations, even along the banks of rivers, were comparatively healthy.

In case of sickness, physicians are to be found in almost every county, and every season adds to their number. Charges are somewhat higher than in the Northern States. Many families keep a few simple articles of medicine, and administer for themselves. Calomel is a specific, and is taken by multitudes, without hesitation or fear of danger. From fifteen to twenty grains are an ordinary dose for a cathartic. Whenever nausea of the stomach, pains in the limbs, and yawning, or a chill, indicate the approach of disease, a dose of calomel is taken at night in a little apple, honey, or other suitable substance, and followed up in the morning with a dose of castor oil, or salts, to produce a brisk purge. Sometimes an emetic is preferred: either a cathartic or an emetic will leave the system under some debility. The mistake frequently made is, in not following up the evacuating medicine with tonics. This should be done, invariably, unless the paroxysm of fever has commenced. A few doses of sulphate of quinine, or Peruvian bark in its crude state, will restore the system to its natural tone. To prevent an attack of fever, medicine should be taken on the very first symptoms of a diseased stomach; it should not be tampered with, but taken in sufficient doses to relieve the system from morbid effects, and then followed up by tonics, to restore its vigor and prevent relapse.

New comers will find it advantageous for protecting themselves from the damp atmosphere at night, to provide close dwellings; yet when the air is clear, to leave open doors and windows at night for free circulation, but not to sleep directly in the current of air; and invariably to wear thin clothing in the heat of the day, and put on thicker garments at night, and in wet and cloudy weather.

I have observed that those families are seldom sick who live in comfortable houses, with tight floors, and well ventilated rooms, and who, upon change of weather, and especially in time of rains, make a little fire in the chimney, although the thermometer might not indicate the necessity.

In fine, I am prepared to give my opinion,

decidedly, in favor of the general health of this country and climate. I would not certainly be answerable for all the bad locations, the imprudences and whims of all classes of emigrants, which may operate unfavorably to health. I only speak for myself and family. I decidedly prefer this climate, with all its miasm, to New England, with its north-east winds, and damp, 'raw,' and pulmonary atmosphere. We very seldom have fogs in Illinois and Missouri. My memoranda, kept with considerable accuracy, for twelve years, give not more than half a dozen foggy mornings in a year.

The following comparisons between St. Louis and several eastern cities, will afford some evidence of the opinions expressed above. I have remarked, already, that 1821 was more sickly in St. Louis than any preceding year, and deaths were more numerous in proportion to the population. Some cases of fever were more malignant in 1820, in that place, but deaths were more frequent the following season. I solemnized the marriage of a young lady of my acquaintance, who was under the age of fourteen years. In eight days she was a widow. At the funeral of a gentleman, the same season, who left a widow under twenty years, there were present thirteen widows, all under twenty-four years of age, and all had lost their companions that season. Young men were victims more than any other age or condition. And yet I am

5

prepared to show, that St. Louis, that summer, was not more sickly than several eastern cities were in 1820 and 1823.

The population of St. Louis, in 1821, varied but little from 5000; the number of deaths during that year was 136. This account was taken by the Rev. Salmon Giddings, who was particular in collecting the facts. The proportion of the deaths to the population, was one to thirty-five.

In 1820, Boston contained a population of 43,893; number of deaths 1103; proportion of one to thirty-nine and three fourths.

New York. Population 123,000; deaths 3515; proportion of one to a fraction less than thirty-five.

Philadelphia. Population 108,000; deaths 3374; proportion of one to thirty-two.

Baltimore. Population 62,000; deaths 1625; proportion of one to thirty-eight.

The aggregate population of these four cities, in 1820, was 336,893; the aggregate number of deaths, 9617; proportion of one to thirty-five, the same as that of St. Louis.

In 1823, Boston contained a population estimated at 45,000; number of deaths, by official returns, 1154; proportion of one to thirty-nine.

New York. Population about 130,000; deaths 3444; proportion of one to thirty-seven and two thirds.

Philadelphia. Population about 120,000;

deaths 4600; proportion of one to twenty-six. [This was an uncommonly sickly season in Philadelphia.]

Baltimore. Population estimated at 65,000; deaths 2108; proportion of one to thirty and two thirds.

I have thus selected the mortality of St. Louis during the most sickly season since my residence in this country, and compared it with the bills of mortality of four eastern cities for two years (those of 1820 and 1823), and the result is favorable to the health of St. Louis, and, by consequence, to the adjoining States. For ten years past, there has been no general sickness in St. Louis, during the summer and autumnal months, excepting the cholera, in 1832.

Some parts of Indiana and Ohio are unquestionably more subject to bilious attacks than Illinois. The reason is obvious. Much of that region is heavily timbered; and, upon cutting it away in spots, and letting in the rays of the sun upon vegetable matter undergoing decomposition, miasmata are generated. These regions will become comparatively healthy, when put under general cultivation.

The story is told, that the late emperor of France lay encamped with one of his armies near a place reputed unhealthy, when one of his officers requested a furlough. The reason being asked, and given,—that the place was unhealthy, and the applicant feared to die an inglorious death, from fever:—Napoleon replied, in his accustomed laconic style, "Go to your post; men die every where."

If a family emigrate to a new and distant country, and any of the number sicken and die, we are apt to indulge in unavailing regret at the removal; whereas, had the same afflictive event happened before removal, it would have been regarded in quite a different light. Let, then, none come to the West, who do not expect to be sick and to die, whenever divine Providence shall see fit so to order events.

The *milk sickness* is a disease of a singular character, which prevails in certain places. It first affects animals, especially cows, and from them is communicated to the human system by eating the milk, or flesh. The symptoms of the disease indicate poison; and the patient is affected nearly in the same way, as when poisonous ingredients have been received into the system. Cattle, when attacked by it, usually die. In many instances it proves mortal in the human system; in others, it yields to the skill of the physician. Much speculation has been had upon its cause, which is still unknown. The prevailing idea is, that it is caused by some poisonous substance eaten by the cattle, but whether vegetable or mineral, remainsun determined. Physicians and others have attempted to ascertain the cause of this disease, but hitherto without success.

It infests only particular spots, or small districts, and these are soon found out. There are places in Ohio, Indiana, and the southern

States, where it exists. Its effects are more frequent in autumn than any other season; and to guard against it, the people either keep their cows in a pasture, or refuse to use their milk. Some have supposed this disease to be produced by the cattle feeding on the *cicuta virosa*, or water hemlock; as a similar disease once infested the cattle in the north of Europe, the cause of which was traced out by the great naturalist Linnæus; but it is not known that this species of plant exists amongst the botanical productions of Missouri and Illinois.

Anxious to furnish all the information, on this very important subject, to persons desirous of emigrating to the West, I will prolong this chapter by inserting the following

"*Advice to Emigrants, recent Settlers, and to those visiting the southern Country.*

"The outlines which have already been given, will afford some information to emigrants from other sections of the Union, or from Europe. We will now offer a few cautionary remarks, particularly intended for such as are about to settle, or have recently settled in this section of the United States.

"Of new comers, there are two tolerably distinct classes: the one comprising farmers, mechanics, and indeed all those who calculate on obtaining a subsistence by manual industry; the other is composed of professional men, tradesmen, and adventurers of every description. Towards the first class our attention is now directed, premising that throughout a great portion of the western country, except in large towns, nearly every mechanic is almost necessarily a farmer; the population being in but few places sufficiently dense to support that designation of mechanical employments which is common in the Eastern and Middle States.

"For the industrious and temperate of this class, our country holds forth inducements, which are not generally known or understood.

"The language of indiscriminate panegyric, which has been bestowed on its climate and soil, has conveyed little information, and is the source of many fears and suspicions in the minds of people at a distance. Other accounts have described the western country as uniformly sickly; but the habit of exaggeration in its favor, has been most prevalent; neither need we wonder, when much of the information communicated, has been afforded by interested landholders, or speculators, and by travelers, whose views have been superficial, and whose journeys have been performed generally, either on the rivers, or by post roads.

"The first inquiry of a substantial farmer, from one of the old settled States is, mostly, for good land, in the vicinity of a market; and, afterwards, whether the situation be healthy. It is true that there are many places in the western country, affording the qualities expressed in this description, but they are perhaps all occupied; and it would be, in several respects, more advisable for a farmer, possessing even a considerable sum of money in hand, to inquire first for a healthy situation, and then good land.

"The spirit of improvement throughout the United States, especially evinced in canalling and rail-roads, will, it is hoped, in a few years, open modes of communication, which, as yet, are wanting with the markets.

"The same remarks will apply to the poorer class of emigrants. If they value their own health, and that of their families, the main object of their attention will be to secure, if possible, a situation remote from the fogs that hover over the channels of large rivers, which become partly dry in summer, and from the neighborhood of swamps, marshes, ponds and small lakes.

"Every person, on coming from beyond the mountains, and especially from the Eastern States, or Europe, will have to undergo some degree of change in his constitution, before it becomes naturalized to the climate; and all who move from a cold to a considerably warmer part of the western country will experience the same alteration; it will, therefore, be wisdom for the individual brought up in a more rigorous climate, that he seek a situation where the circulation of air is unimpeded and free, and that he avoid those flat and marshy districts, which have been already described.

"Those who settle in new countries are

almost universally exposed to inconveniences which have an unfavorable influence on health. They are seldom able, for a length of time, to erect comfortable places of residence; and indeed, many postpone this important object of attention, even after their circumstances will permit them to build comfortable dwelling-houses.

"Wool is mostly a scarce article in new settlements, so that cotton and linen garments are too frequently worn in winter. There is another circumstance, which no doubt has an unfavorable influence on health, especially among the poorer class: it is the want, during the summer season particularly, of substantial food. This is sometimes owing to indolence, or improvidence; but perhaps oftener to the circumstances in which a few families are placed, at a distance from any established or opulent settlement.

"Erroneous views are too generally entertained in relation to hardening the human system; and the analogies drawn from savage life, are altogether inconclusive. The manners of the North American Indians are essentially different from those of the whites. It is true, there is a portion of the latter, especially in Illinois and Missouri, who from infancy are educated almost in the habits of the aborigines.

"We have frequently heard the example of savages referred to, as an argument in favor of attempting to strengthen the constitution by exposure.* There is plausibility in this; but might not the example of the negroes in the lower parts of South Carolina and

* Uniform exposure to the weather is favorable to health. I can affirm this from long experience and observation. Our hunters and surveyors, who uniformly spend their time for weeks in the woods and prairies, who wade in the water, swim creeks, are drenched in the rains and dews, and sleep in the open air or a camp at night, very rarely are attacked with fevers. I have known repeated instances of young men, brought up delicately in the eastern cities, accustomed, as clerks, to a sedentary life, with feeble constitutions,—I have known such, repeatedly, to enter upon the business of surveying the public lands, or in the hunting and trapping business, be absent for months, and return with robust health. It is a common thing for a frontier man, whose health is on the decline, and especially when indications of pulmonary affection appear, to engage in a hunting expedition to renovate his health. I state these facts, and leave it to the medical faculty to explain the *why* and *wherefore*. One circumstance may deserve attention. All these men, as do the Indians, *sleep with their feet towards the fire at night.* And it is a common notion with this class, that if the feet are kept hot through the night, however cold the atmosphere, or however much exposed the rest of the body, no evil consequences will ensue. I have passed many a night in this position, after fatiguing rides of thirty or forty miles in the day on our extreme frontiers, and through rains, and never experienced any inconvenience to health, if I could get a pallet on the cabin floor, and my feet to the fire.

Those who are exposed to these hardships but occasionally, when compelled by necessity, and who endeavor to protect themselves at all other times, usually suffer after such exposure.

I have observed that children, when left to run in the open air and weather, who go barefoot, and oftentimes with a single light garment around them, who sleep on the floor at night, are more healthy than those who are protected.

Georgia, be also quoted as evidencing the propriety of living on corn-meal and sweet potatoes, and working every day in the water of a rice-field, during the sickly season? They are, generally, more healthy than the whites who own them, and who reside on the plantations in the summer. The civilized man may turn to savage life perhaps with safety, as regards health; but then he must plunge with the Indian into the depths of the forest, and observe consistency in all his habits. These pages are not written, however, for such as are disposed to consider themselves beyond the pale of civilized society; but for the reflecting part of the community, who can estimate the advantages to be derived from a prudent care of health.

"Much disease, especially in the more recently settled parts of this country, is consequent to neglecting simple and comfortable precautionary means: sometimes, this neglect is owing to misdirected industry, and at others to laziness, or evil habits.

"To have a dry house, if it be a log one, with the openings between the logs well filled up, so that it may be kept warm in winter; to fill up all the holes in its vicinity which may contain stagnant water; to have a good clean spring or well, sufficient clothing, and a reasonable supply of provisions, should be the first object of a settler's attention. But frequently a little, wet, smoky cabin or hovel is erected, with the floor scarcely separated from the ground, and admitting the damp and unwholesome air: all hands, that can work, are impelled, by the father's example, to labor beyond their strength, and more land is cleared and planted with corn than is well tended; for over-exertion, change in the manner of living, and the influence of other debilitating causes, which have been mentioned, bring sickness on at least a part of the family, before the summer is half over.

"It is unnecessary for even the poorest emigrant to encounter these causes of distress, unless seduced by the misrepresentations of some interested landholder, or by the fantasies of his own brain, to an unhealthy and desolate situation, where he can neither help himself, nor be assisted by others.

"Many persons, on moving into the *back woods*, who have been accustomed to the decencies of life, think it little matter how they live, because *no one sees them.* Thus we have known a family of some opulence to reside for years in a cabin unfit for the abode of any human being, because they could not find time to build a house! and whenever it rained hard, the females were necessarily engaged in rolling the beds from one corner of the room to another, in order to save them from the water that poured in through the roof. This cabin was intended at first as only a very temporary residence, and was erected on the edge of a swamp, for the convenience of being near a spring. How unreasonable must such people be, if they expect health!

"Clothing for winter, should be prepared in summer. It is a common, but very incorrect practice among many farmers, both west and east of the Alleghany mountains, to postpone wearing winter clothing until the weather has become extremely cold: this is a fruitful source of pulmonary diseases, of rheumatisms and of fevers.

"With regard to providing a sufficiency of nourishing food, no specific directions can be given, further than to recommend,—what is much neglected,—particular attention to a good garden spot; and to remark, that those who devote undivided attention to cultivating the soil, receive more uniform supplies of suitable nourishment than the more indolent, who spend a considerable portion of their time in hunting.

"New settlers are not unfrequently troubled with diseases of the skin, which are often supposed to be the itch: for these eruptions they generally use repellent external applications: this plan of treatment is prejudicial.

"The most proper time for the removal of families to this country, from the Atlantic States, is early in the spring, while the rivers are full; or, if the journey be made by land, as soon as the roads are sufficiently settled, and the waters abated.

"Persons unaccustomed to the climate of the Lower Mississippi country, are necessarily exposed, whilst there in the summer season, to many causes of disease. It will be advisable for such to have a prudent care of their health, and yet a care distinct from that finical timidity, which renders them liable to early attacks of sickness.

"There is one important consideration, which perhaps has been somewhat overlooked by medical men, who have written on this subject. Natives of colder and healthier regions, when exposed in southern and sickly climates, experience, if they remain any length of time without evident and violent disease, an alteration in the condition of the liver, and of the secreted bile itself; when it passes through the bowels, its color being much darker than usual. Sometimes, indeed, it appears to be 'locked up in the liver,' the stools having an ashen appearance. This state of the biliary secretion is frequently accompanied, although the patient is otherwise apparently in tolerable health, by a pain over the eye-balls, particularly when the eyes are rolled upward.

"The proper mode of treatment for such symptoms is, to take, without delay, not less than twenty grains of calomel, and, in eight hours, a wine-glass full of castor oil. The tone of the stomach should not be suffered to sink too much after the operation of the medicine, which, if necessary, may be repeated in twenty-four hours. Sulphate of quinine, or other tonics, with nutritive food, which is easy of digestion, should also be taken in moderate portions at a time.

"Where diseases are rapid in their progress, and dangerous, no time is to be lost. The practice of taking salts and other aperients, when in exposed situations, and for the purpose of preventing disease, is injurious. It is sufficient, that the bowels be kept in a natural and healthy state; for all cathartics, even the mildest, have a tendency to nauseate the stomach, create debility, and weaken the digestive faculty. A reduction of tone in the system, which is always advantageous, will be more safely effected by using somewhat less than usual of animal food, and of spirituous, strong, vinous or fermented liquors. The robust will derive benefit from losing a little blood.

"It ought to be well understood, that as we approximate tropical climates, the doses of medicine, when taken, should be increased in quantity, and repeated with less delay than is admissible in colder countries. Exposure to the night air is certainly prejudicial; so also is the intense heat of the sun, in the middle of the day. Violent exercise should also be avoided. Bathing, daily, in water of a comfortable temperature, is a very commendable practice; and cotton, worn next the skin, is preferable to linen.

"It is impossible to prevent the influence of an atmosphere pregnant with the causes of disease; but the operation of those causes may generally be counteracted, by attention to the rules laid down; and it is no small consolation to be aware that, on recovery from the first attack, the system is better adapted to meet and sustain a second of a similar nature. The reader will understand that we do not allude to relapses, occurring while the system is enfeebled by the consequences of disease."

To the foregoing remarks, I add the following, from an address of Judge Hall, December 10, 1827:

"The climate, particularly in reference to its influence on the human system, presents another subject of investigation. The western country has been considered unhealthy, and there have been writers, whose disturbed imaginations have misled them into a belief that the whole land was continually exposed to the most awful visitations of Providence, among which have been numbered the hurricane, the pestilence, and the earthquake. If we have been content to smile at such exaggeration, while few had leisure to attempt a serious refutation, and while the facts upon which any deliberate opinion must have been based, had not been sufficiently tested by experience, the time has now arrived when it is no longer excusable to submit in silence to the reproaches of ignorance or malice. It is proper, however, to remark, as well in extenuation of those who have assailed our country, as in the support of the confidential denial, which I feel authorized to make to their assertions, that a vast improvement in the article of health has taken place within a few years. Diseases are now mild which were once malignant, and their occurrence is annually becoming less frequent. This happy change affords strong authority for the belief, that, although the maladies which have heretofore afflicted us were partly imputable to the climate, other and more powerful causes of disease must have existed, which have vanished. We, who came to the frontier while the axe was still busy in the forest, and when thousands of the acres which now yield abundance to the farmer, were unreclaimed and tenantless, have seen the existence of our fellow-citizens assailed by other than the ordinary ministers of death. Toil, privation and exposure, have hurried many to the grave; imprudence and carelessness of life, have sent crowds of victims prematurely to the tomb. It is not to be denied that the margins of our great streams in general, and many spots in the vicinity of extensive marshes, are subject to bilious diseases; but it may be as confidently asserted, that the interior country is healthy. Yet the first settlers invariably selected the rich alluvion lands upon the navigable rivers, in preference to the scarcely less fertile soil of the prairies, lying in situations less accessible and more remote from market. They came to a wilderness in which houses were not prepared for their reception, nor food, other than that supplied by nature, provided for their sustenance. They often encamped on the margin of the river, exposed to its chilly atmosphere, without a tent to shelter, with scarcely a blanket to protect them. Their first habitations were rude cabins, affording scarcely a shelter from the rain, and too frail to afford protection from the burning heat of the noonday sun, or the chilling effects of the midnight blast. As their families increased, another and another cabin was added, as crazy and as cheerless as the first, until admonished of the increase of their own substance, the influx of wealthier neighbors, and the general improvement of the country around them, they were allured by pride to do that to which they never would nave been impelled by suffering. The gratuitous exposure to the climate, which the backwoodsman seems rather to court than avoid, is a subject of common remark. No extremity of weather confines him to the shelter of his own roof. Whether the object be business or pleasure, it is pursued with the same composure, amid the shadows of the night, or the howling of the tempest, as in the most genial season. Nor is this trait of character confined to woodsmen or to farmers: examples of hardihood are contagious, and in this country all ranks of people neglect or despise the ordinary precautions with respect to health. Judges and lawyers, merchants, physicians and ministers of the gospel, set the seasons at defiance, in the pursuit of their respective callings. They prosecute their journeys, regardless of weather; and learn at last to feel little inconvenience from the exposure, which is silently undermining their constitutions. Is it extraordinary, that people thus exposed should be attacked by violent maladies? Would it not be more wonderful that such a careless prodigality of life could pass with impunity? These remarks might be extended; the food of the first settler, consisting chiefly of fresh meat, without vegetables, and often without salt; the common use of ardent spirits, the want of medical aid, by which diseases, at first simple, being neglected, become dangerous; and other evils peculiar to a new country, might be noticed, as fruitful sources of disease; but I have already dwelt sufficiently on this subject. That this country is decidedly healthy, I feel no hesitation in declaring; but neither argument nor naked assertions will convince the world. Let us collect such facts as amount to evidence, and establish the truth, by undeniable demonstration."

CHAPTER IV.

CHARACTER, MANNERS, AND PURSUITS OF THE PEOPLE.

Cotton and Sugar Planters—Farmers—Population of the large Towns and Cities—Frontier Class—Hunters and Trappers—Boatmen.

THERE is great diversity in the character and habits of the population of the Valley of the Mississippi.

Those who have emigrated from the Atlantic States, as have a very large proportion of those persons who were not born in the Valley, of course do not differ essentially from the remaining population of those States. Some slight shades of difference are perceptible in such persons as have lived long enough in the country to become assimilated to the habits, and partake of the feelings, of western people.

Emigrants from Europe have brought the peculiarities of the nations and countries from whence they have originated, but are fast losing their national manners and feelings, and, to use a provincial term, will soon become "westernized."

The march of emigration from the Atlantic border has been nearly in a line due west. Tennessee was settled by Carolinians, and Kentucky by Virginians. Ohio received the basis of its population from the States in the same parallel, and hence partakes of all the varieties from Maryland to New England. Michigan is substantially a child of New York. The planters of the south have gone to Mississippi, Louisiana, and the southern part of Arkansas. Kentucky and Tennessee have spread their sons and daughters over Indiana, Illinois and Missouri; but the two former States are now receiving great numbers of emigrants from all the northern States including Ohio, and multitudes from the south, who desire to remove beyond the boundaries and influence of a slave population.

Slavery in the West keeps nearly in the same parallels as it holds in the East, and is receding south, as it does on the Atlantic coast. Many descendants of the Scotch, Irish and Germans, have come into the frontier States from Western Pennsylvania.

We have European emigrants from Great Britain and Ireland. Those of the latter are more generally found about our large towns and cities, and along the lines of canalling.

The French were the explorers and early settlers of the Valley immediately bordering on the Mississippi, 150 years since. They formed the basis of population of Louisiana a few years since, but are relatively diminishing before the emigration from other States of the Union. Their descendants show many of the peculiar and distinctive traits of that people in all countries. They possess mild vivacity, and gayety, and are distinguished for their quiet, inoffensive, domestic, frugal, and unenterprising spirit and manners. The poorer class of French are rather peculiar and unique. Their ancestors were isolated from the rest of the world, had no object of excitement or ambition, cared little for wealth, or the accumulation of property, and were accustomed to hunt, make voyages in their canoes. smoke, and traffic with the Indians. But few of them knew how to read and write. Accustomed from infancy to the life of huntsmen, trappers and boatmen, they make but indifferent farmers. They are contented to live in the same rude, but neatly whitewashed cabin, cultivate the same corn-fields in the same mode, and drive the same rudely constructed horse-cart their fathers did. In the neatness of their gardens, which are usually cultivated by the females, they excel the Americans. They are the *coureurs du bois* of the West.

The European Germans are now coming into the Valley by thousands, and, for a time, will retain their manners and language.

Cotton and Sugar Planters. These people, found chiefly in Mississippi, Louisiana, and the southern part of Arkansas, have a great degree of similarity. They are noted for their high-mindedness, generosity, liberality, hospitality, sociability quick sense of honor, resentment of injuries, indolence, and, in too many cases, dissipation. They are much addicted to the sports of the turf and the vices of the gaming table. Still there are many planters of strictly moral, and even religious habits. They are excessively jealous of their political rights, yet frank and open-hearted in their dispositions, and carry the duties of hospitality to a great extent. Having overseers on most of their plantations, the labor being performed by the slaves, they have much leisure, and are averse to much personal attention to business. They dislike care, profound thinking and deep impressions. The young men are volatile, gay, dashing and reckless spirits, fond of excitement and high life. There is a fatal propensity amongst the southern planters to decide quarrels, and even trivial disputes, by duels. But there are also many amiable and noble traits of character amongst this class; and if the principles of the Bible and religion could be brought to exert a controlling influence, there would be a noble spirited race of peoplein the south-western States.

It cannot be expected that I should pass in entire silence the system of slave-holding in the Lower Valley, or its influence on the manners and habits of the people. This state of society seems unavoidable at present, though I have no idea or expectation it will be perpetual. Opposite sentiments and feelings are spreading over the whole earth, and a person must have been a very inattentive observer of the tendencies and effects of the diffusion of liberal principles, not to perceive that hereditary, domestic servitude must have an end.

This is a subject, however, that, from our civil compact, belongs exclusively to the citizens of the States concerned; and if not unreasonably annoyed, the farming slave-holding States, as Kentucky, Tennesse, and Missouri, will soon provide for its eventual termination. Doubtless, in the cotton and sugar-growing States it will retain its hold with more tenacity; but the influence of free principles will roll onward until the evil is annihilated.

The barbarous and unwise regulations in some of the planting States, *which prohibit the slaves from being taught to read*, are a serious impediment to the moral and religious instruction of that numerous and unfortunate class. Such laws display, on the part of the law makers, little knowledge of human nature, and the real tendency of things. To keep *slaves* entirely ignorant of the rights of man, in this spirit-stirring age, is utterly impossible. Seek out the remotest and darkest corner of Louisiana, and plant every guard that is possible around the negro quarters, and the light of truth will penetrate. Slaves will find out, for they already know it, that they possess rights as men. And here is the fatal mistake now committed in the southern slave-holding States —legislating against the instruction of their slaves, to keep them from knowing their rights. They will obtain some loose, vague, and undefined notion of the doctrine of human rights, and the unrighteousness of oppression in this republican country. Being kept from all the moral and religious instruction which Sabbath schools, the Bible, and other good books are calculated to impart, and with those undefined notions of liberty, and without any moral principle, they are prepared to enter into the first insurrectionary movement proposed by some artful and talented leader. The same notion prevailed in the West Indies half a century since, and many of the planters resisted and persecuted the benevolent Moravians, who went there to instruct the blacks in the principles and duties of religion. A few of the planters reasoned justly. They invited these benevolent men on their plantations, and gave them full liberty on the Sabbath, and at other suitable seasons, to instruct their slaves. The happiest effects followed. On these plantations, where riot, misrule, and threatened insurrections, had once spread a panic through the colony, order, quietness and submission followed. Such would be the effects if the southern planter would invite the minister of the gospel and the Sabbath school teacher to visit his plantation, allow his slaves to be instructed to read, and each to be furnished with a copy of the Scriptures. The southern planter hourly lives under the most terrific apprehensions. It is in vain to disguise the fact. As Mr. Randolph once significantly said in Congress, "*when the night bell rings, the mother hugs her infant closer to her breast.*" Slavery, under any circumstances, is a bitter draught—equally bitter to him who tenders the cup, and to him who drinks it. But in all the northern slave-holding States, it is comparatively mild. Its condition would be much alleviated, and the planter might sleep securely, if he would abolish his barbarous laws, more congenial with Asiatic despotism than American republicanism, and provide for his slaves the benefits of wholesome instruction. Philanthropy and interest unite in their demands upon every southern planter to provide Sunday school instruction for his slaves.

The planting region of the Lower Valley furnishes an immense market for the productions and manufactures of the Upper Valley. Indirectly, the Louisiana sugar business is a source of profit to the farmer of Illinois and Missouri. Pork, beef, corn, corn-meal, flour, potatoes, butter, hay, &c., in vast quantities, go to supply these plantations. In laying in their stores, the sugar planters usually purchase one barrel of second or third quality of beef or pork per annum, for each laborer. Large drafts for sugar-mills, engines and boilers, are made upon the Cincinnati and Pittsburgh iron foundries. Mules and horses are driven from the upper country, or from the Mexican dominions, to keep up the supply.

The commerce of the upper country that concentrates at New Orleans is amazing, and every year is rapidly increasing. Sixteen hundred arrivals of steam-boats took place in 1832, and the estimated number, in 1835, is 2300.

Farmers. In the northern half of the Valley the productions, and the modes of cultivation and living are such as to characterize a large proportion of the population as farmers. No country on earth has such facilities for agriculture. The soil is abundantly fertile, the seasons ordinarily favorable to the growth and maturity of crops, and every farmer, in a few years, with reasonable industry, becomes comparatively independent. Tobacco and hemp are among the staple productions of Kentucky.

Neat cattle, horses, mules and swine are its stock. Some stock growers have monopolized the smaller farms till they are surrounded with several thousand acres. Blue grass pastures furnish summer feed, and extensive fields of corn, cut up near the ground, and stacked in the fields, furnish stores for fattening stock in the winter.

In some counties, raising of stock has taken place of all other business. The Scioto Valley, and other districts in Ohio, are famous for fine, well-fed beef. Thousands of young cattle are purchased by the Ohio graziers, at the close of winter, of the farmers of Illinois and Missouri. The Miami and White-water sections of Ohio and Indiana, abound with swine. Cincinnati has been the great pork mart of the world. 150,000 head of hogs have been frequently slaughtered there in a season. About 75,000 is estimated to have been the number slaughtered at that place the past season. This apparent falling off in the pork business, at Cincinnati, is accounted for by the vast increase of business at other places. Since the opening of the canals in Ohio, many provision establishments have been made along their line. Much business of the kind is now done at Terre Haute, and other towns on the Wabash,—at Madison, Louisville, and other towns on the Ohio,—at Alton, and other places in Illinois.

The farmers of the West are independent in feeling, plain in dress, simple in manners, frank and hospitable in their dwellings, and soon acquire a competency by moderate labor. Those from Kentucky, Tennessee, or other States south of the Ohio river, have large fields, well cultivated, and enclosed with strong built rail or worm fences, but they often neglect to provide spacious barns and other out-houses for their grain, hay and stock. The influence of habit is powerful. A Kentuckian would look with contempt upon the low fences of a New Englander, as indicating thriftless habits, while the latter would point at the unsheltered stacks of wheat, and dirty threshing-floor of the former, as proof direct of bad economy and wastefulness.

Population of the Cities and large Towns. The population of western towns does not differ essentially from the same class in the Atlantic States, excepting there is much less division into grades and ranks, less ignorance, low depravity and squalid poverty amongst the poor, and less aristocratic feeling amongst the rich. As there is never any lack of employment for laborers of every description, there is comparatively no suffering from that cause. And the hospitable habits of the people provide for the sick, infirm and helpless. Doubtless, our *circumstances*, more than any thing else, cause these shades of difference. The common mechanic is on a social equality with the merchant, the lawyer, the physician, and the minister. They have shared in the same fatigues and privations, partook of the same homely fare, in many instances have fought side by side in defence of their homes against the inroads of savages,—are frequently elected to the same posts of honor, and have accumulated property simultaneously. Many mechanics in the western cities and towns, are the owners of their own dwellings, and of other buildings, which they rent. I have known many a wealthy merchant, or professional gentleman occupy on rent, a building worth several thousand dollars, the property of some industrious mechanic, who, but a few years previous, was an apprentice lad, or worked at his trade as a journeyman.

Any sober, industrious mechanic can place himself in affluent circumstances, and place his children on an equality with the children of the commercial and professional community, by migrating to any of our new and rising western towns. They will find no occasion here for combinations to sustain their interests, nor meet with annoyance from gangs of unprincipled foreigners, under the imposing names of "Trades Unions."

Manufactures of various kinds are carried on in our western cities. Pittsburgh has been characterized as the "Birmingham of America." The manufactures of iron, machinery, and glass, and the building of steam-boats, are carried on to a great extent.

Iron and salt are made in great quantities, in Western Pennsylvania and Western Virginia. Steam-boats are built, to a considerable extent, at Fulton, two miles above Cincinnati, and occasionally, at many other places on the Ohio and Mississippi rivers. Alton offers great facilities for this business. Cotton bagging, bale ropes and cordage, are manufactured in Tennessee and Kentucky. The following article, from the Covington Enquirer, gives a few items of the industry and enterprise of Kentucky,—of the manufacture of Newport and Covington. Both of these thriving towns lie at the mouth of the Licking river, the one on the right bank, and the other on the left, and both in direct view of Cincinnati.

Manufactures in Covington and Newport.

"Founding the calculation upon the actual manufactures of October, and the known power of their machinery, the company will, the ensuing year, give employment to more than four hundred operatives, and manufacture,—

Cotton bagging,	60,000	pounds.
" yarns,	84,000	"
Bale rope,	274,268	"
Cordage,	448,000	"
Linseys,	44,592	yards,
Cotton plains,	63,588	"
Kentucky jeans,	97,344	"
Cotton bagging and hemp,	548,530	"

"Estimating bale rope and cotton bagging at thirty-three per cent. under the price at which the company have sold these articles for the last six months, the manufactures of this company, during the ensuing year, will amount to $358,548 44. Almost all the manufactures at Covington and Newport being exported to foreign markets, it will result that the annual exports from these points will, in round numbers, be, from the

Interior,	$750,000
Campbell county,	150,000
Boone "	234,000
Covington,	548,500
Newport,	358,500
	$2,041,000

"The Newport Manufacturing Company has depended principally, for its supply of hemp, on the production of Mason county, of which Maysville is the market. This season, they have not been able to get a supply at Maysville; and it is a remarkable fact in the history of the hemp manufactories in Kentucky, that this company, owing to the scarcity and high prices of hemp in Kentucky, *have imported this season, 354,201 lbs. Russia hemp.*"

Various manufactories are springing up in all the new States, which will be noticed under their proper heads.

The number of merchants and traders is very great in the Valley of the Mississippi, yet mercantile business is rapidly increasing. Thousands of the farmers of the West are partial traders. They take their own produce, in their own flat boats, down the rivers to the market of the lower country.

Frontier Class of Population. The rough, sturdy habits of the backwoodsmen, living in that plenty which depends on God and nature, have laid the foundation of independent thought and feeling deep in the minds of western people.

Generally, in all the western settlements, three classes, like the waves of the ocean, have rolled one after the other. First, comes the pioneer, who depends for the subsistence of his family chiefly upon the natural growth of vegetation, called the "range," and the proceeds of hunting. His implements of agriculture are rude, chiefly of his own make, and his efforts directed mainly to a crop of corn, and a "truck patch." The last is a rude garden for growing cabbage, beans, corn for roasting ears, cucumbers and potatoes. A log cabin, and, occasionally, a stable and corn-crib, and a field of a dozen acres, the timber girdled or "deadened," and fenced, are enough for his occupancy. It is quite immaterial whether he ever becomes the owner of the soil. He is the occupant for the time being, pays no rent, and feels as independent as the "lord of the manor." With a horse, cow, and one or two breeders of swine, he strikes into the woods with his family, and becomes the founder of a new county, or perhaps State. He builds his cabin, gathers around him a few other families of similar taste and habits, and occupies till the range is somewhat subdued, and hunting a little precarious, or, which is more frequently the case, till neighbors crowd around, roads, bridges and fields annoy him, and he lacks elbow room. The preëmption law enables him to dispose of his cabin and corn-field, to the next class of emigrants, and, to employ his own figures, he "breaks for the high timber," "clears out for the New Purchase," or migrates to Arkansas, or Texas, to work the same process over.

The next class of emigrants purchase the lands, add field to field, clear out the roads, throw rough bridges over the streams, put up hewn log houses, with glass windows, and brick or stone chimneys, occasionally plant orchards, build mills, school-houses, court-houses, &c., and exhibit the picture and forms of plain, frugal, civilized life.

Another wave rolls on. The men of capital and enterprise come. The "settler" is ready to sell out, and take the advantage of the rise of property,—push farther into the interior, and become himself, a man of capital and enterprise in turn. The small village rises to a spacious town or city; substantial edifices of brick, extensive fields, orchards, gardens, colleges and churches are seen. Broadcloths, silks, leghorns, crapes, and all the refinements luxuries, elegancies, frivolities and fashions, are in vogue. Thus wave after wave is rolling westward:—the real *el dorado* is still farther on.

A portion of the two first classes remain stationary amidst the general movement, improve their habits and condition, and rise in the scale of society.

The writer has traveled much amongst the first class,—the real pioneers. He has lived many years in connexion with the second grade; and now the third wave is sweeping over large districts of Indiana, Illinois and Missouri. Migration has become almost a habit, in the West. Hundreds of men can be found, not fifty years of age, who have settled for the fourth, fifth, or sixth time on a new spot. To sell out, and remove only a few hundred miles, makes up a portion of the variety of backwoods life and manners.

But to return to the frontier class:

1. *Dress.* The hunting-shirt is universally worn. This is a kind of loose, open frock, reaching half way down the thighs, with large sleeves, the body open in front, lapped over and belted with a leathern girdle, held together by a buckle. The cape is large, and usually fringed with different colored cloth from that of the body. The bosom of this dress sometimes serves as a wallet for a "chunk" of bread, jerk or smoke-dried venison, and other articles. It is made either of dressed deer-skins, linsey, coarse linen, or cotton. The shirt, waistcoat and pantaloons are of similar articles, and of the customary form. Wrappers, of cloth or dressed skins, called "leggins," are tied round the legs when traveling. Moccasins, of deer skins, shoe-packs and rough shoes, the leather tanned and cobbled by the owner, are worn on the feet.

The females dress in a coarse gown, of cotton, a bonnet of the same stuff, and denominated in the Eastern States a "sun bonnet." The latter is constantly worn through the day, especially when company is present. The clothing, for both sexes, is made at home. The wheel and loom are common articles of furniture, in every cabin.

2. *Dwellings.* "Cabin" is the name for a plain, rough log house, throughout the West. The spot being selected, usually, in the timbered land, and near some spring, the first

operation of the newly arrived emigrant is to cut about forty logs, of the proper size and length, for a single cabin, or twice that number for a double one, and haul them to the spot. A large oak, or other suitable timber, of staight grain and free from limbs, is selected for clapboards for the roof: these are four feet in length, split with a froe, six or eight inches wide and half an inch thick. *Puncheons* are used for the floor: these are made by splitting trees, about eighteen inches in diameter, into slabs, two or three inches in thickness, and hewn on the upper surface. The door-way is made by cutting out the logs, after raising, of a suitable width, and putting upright pieces of timber at the sides. The shutter is made of clapboards, pinned on cross pieces, hung by wooden hinges, and fastened by a wooden latch. A similar aperture, but wider, is made at one end, for the chimney. The men of the settlement, when notified, collect and raise the building. Four stout men, with axes, are placed on the corners, to notch the logs together, while the rest of the company lift them up. After the roof is on, the body of the building is slightly hewed down, both outside and inside. The roof is formed by shortening each end log, in succession, till one log forms the comb of the roof. The clapboards are put on so as to cover all cracks, and held down by poles, or small logs. The chimney is built of sticks of wood, the largest at the bottom, and the smallest at the top, and laid up with a supply of mud or clay mortar. The interstices between the logs are chinked with strips of wood, and daubed with mortar, both outside and inside. A double cabin consists of two such buildings, with a space of ten or twelve feet between, over which the roof extends.

A *log house*, in western parlance, differs from a cabin, in the logs being hewn on two sides, to an equal thickness, before raising; in having a framed and shingled roof, a brick or stone chimney, windows, tight floors, and are frequently clapboarded on the outside, and plastered within.

A log house thus finished, costs more than a framed one. Cabins are often the temporary dwelling of opulent and highly respectable families.

The axe, auger, froe, drawing-knife, broad axe, and cross-cut saw, are the only tools required in constructing these rude edifices; sometimes the axe and auger only are employed. Not a nail or pane of glass is needed. Cabins are by no means so wretched for residences, as their name imports. They are often roomy, comfortable and neat. If one is not sufficient to accommodate the family, another is added, and another, until sufficient room is obtained.

3. *Furniture, and mode of living.* The genuine backwoodsman makes himself and family comfortable and contented, where those, unaccustomed to his mode of life, would live in unavailing regret, or make a thousand awk-

ward apologies on the visit of a neighbor or traveler. A table is made of a split slab, and supported by four round legs. Clapboards, supported by pins stuck in the logs, answer for shelves for table furniture. The bedstead is often made in the corner of the room, by sticks, placed in the logs, supported at the outward corner by a post, on which clapboards are laid, the ends of which enter the wall, between the logs, and which support the bedding. On the arrival of travelers or visiters, the bed clothing is shared with them, being spread on the puncheon floor, that the feet may project towards the fire. Many a night has the writer passed in this manner, after a fatiguing day's ride, and reposed more comfortably than on a bed of down, in a spacious mansion. All the family, of both sexes, with all the strangers who arrive, often lodge in the same room. In that case, the under garments are never taken off, and no consciousness of impropriety or indelicacy of feeling is manifested. A few pins, stuck in the wall of the cabin, display the dresses of the women and the hunting-shirts of the men. Two small forks, or buck's horns, fastened to a joist, are indispensable articles for the support of the rifle. A loose floor of clapboards, and supported by round poles, is thrown over head, for a loft, which furnishes a place to throw any articles not immediately wanted, and is frequently used for a lodging place for the younger branches of the family. A ladder planted in the corner, behind the door, answers the purpose of stairs.

The necessary table and kitchen furniture are a few pewter dishes and spoons, knives and forks (for which, however, the common hunting-knife is often a substitute), tin cups, for coffee or milk, a water-pail, and a small gourd or calabash for water, with a pot, and iron Dutch oven, constitute the chief articles. Add to these a tray, for wetting up meal, for corn-bread, a coffee-pot and set of cups and saucers, a set of common plates, and the cabin is furnished. The hominy mortar and hand-mill, are in use in all frontier settlements. The first, consists of a block of wood, with an excavation burned at one end, and scraped out with an iron tool, wide at top and narrow at the bottom, that the action of the pestle may operate to the best advantage. Sometimes a stump of a large tree is excavated, while in its natural position; an elastic pole, twenty or thirty feet in length, with the large end fastened under the ground log of the cabin, and the other elevated ten or fifteen feet, and supported by two forks, to which a pestle five or six inches in diameter, and eight or ten feet long, is fixed, on the elevated end, by a large mortice, and a pin put through its lower end, so that two persons can work it in conjunction. This is much used for pounding corn. A very simple instrument, to answer the same purpose, is a circular piece of tin, perforated, and attached to a piece of wood, like a grater, on which the ears of corn are rubbed, for meal. The hand-mill is in the same form as that used in Judea, in the time of our Savior. Two circular stones, about eighteen inches in diameter, constructed like ordinary mill-stones, with a staff let into the runner or upper stone near its outer edge, with the upper end inserted in a joist or board over head, and turned by the hands of two persons, while one feeds it with corn. Horse-mills follow the mortar and hand-mill, in the scale of improvement. They are constructed variously. A *band*-mill is the most simple. A large, upright post is placed on a gudgeon, with shafts extending horizontally, fifteen or twenty feet; around the ends of these is a band of raw hide, twisted, which passes round the trundle head and turns the spindle, and communicates motion to the stone. A *cog*-mill is formed by constructing a rim, with cogs upon the shafts, and a trundle head to correspond. Each person furnishes his own horses to turn the mill, performs his own grinding, and pays toll to the owner, for the use of the mill. Mills, with the wheel on an inclined plane, and carried by oxen standing on the wheel, are much in use in those sections where water-power is not convenient; but these indicate an advance to the second grade of society.

Instead of bolting-cloths, the frontier people use a sieve, or, as it is called here, a "search." This is made from a deer-skin, prepared to resemble parchment, stretched on a hoop, and perforated full of holes, with a hot wire.

Every backwoodsman carries, on all occasions, the means of furnishing his meat. The rifle, bullet-pouch, and horn, hunting-knife, horse and dog, are his constant companions, when from home, and wo be to the wolf, bear, deer or turkey that comes within one hundred and fifty yards of his trail.

With the first emigration there are few mechanics; hence every settler becomes expert in supplying his own necessaries. Besides clearing land, building cabins, and making fences, he stocks his own plough, repairs his wagon and his harness, tans his own leather, makes his shoes, tables, bedsteads, stools, or seats, trays, and a hundred other articles. These may be rudely constructed, but they answer his purpose very well.

The following extracts, from the graphic "SKETCHES OF THE WEST," by James Hall, Esq., completes this extended picture of backwoods manners:

"The traveler, accustomed to different modes of life, is struck with the rude and uncomfortable appearance of every thing about this people; the rudeness of their habitations, the carelessness of their agriculture, the unsightly coarseness of all their implements and furniture, the unambitious homeliness of all their goods and chattels, except the axe, the rifle, and the horse, these being invariably the best and handsomest which their means enable them to procure. But he is mistaken in supposing them indolent or improvident; and is little aware how much ingenuity and toil have been exerted in procuring the few comforts which they possess, in a country without arts, mechanics, money, or commercial intercourse.

"The backwoodsman has many substantial enjoyments. After the fatigue of his journey, and a short season of privation and danger, he finds himself surrounded with plenty. His cattle, hogs and poultry, supply his table with meat; the forest abounds in game; the fertile soil yields abundant crops; he has, of course, bread, milk and butter; the rivers furnish fish, and the woods honey. For these various articles, there is, at first, no market, and the farmer acquires the generous habit of spreading them profusely on his table, and giving them freely to a hungry traveler or indigent neighbor.

"Hospitality and kindness are among the virtues of the first settlers. Exposed to common dangers and toils, they become united by the closest ties of social intercourse. Accustomed to arm in each other's defence, to aid in each other's labor, to assist in the affectionate duty of nursing the sick, and the mournful office of burying the dead, the best affections of the heart are kept in constant exercise; and there is, perhaps, no class of men in our country, who obey the calls of benevolence, with such cheerful promptness, or with so liberal a sacrifice of personal convenience.

"We read marvellous stories of the ferocity of western men. The name of Kentuckian is constantly associated with the idea of fighting, dirking and gouging. The people of whom we are now writing, do not deserve this character. They live together in great harmony, with little contention, and less litigation. The backwoodsmen are a generous and placable race. They are bold and impetuous; and, when differences do arise among them, they are more apt to give vent to their resentment at once, than to brood over their wrongs, or to seek legal redress. But this conduct is productive of harmony; for men are always more guarded in their deportment to each other, and more cautious of giving offence, when they know that the insult will be quickly felt, and instantly resented, than when the consequences of an offensive action are doubtful, and the retaliation distant. We have no evidence that the pioneers of Kentucky were quarrelsome or cruel; and an intimate acquaintance with the same race, at a later period, has led the writer to the conclusion, that they are a humane people; bold and daring, when opposed to an enemy, but amiable in their intercourse with each other and with strangers, and habitually inclined to peace."

In morals, and the essential principles of religion, this class of people are by no means so defective as many imagine. The writer has, repeatedly, been in settlements and districts beyond the pale of civil and criminal law, where the people are "a law unto themselves," where courts, lawyers, sheriffs and constables existed not, and yet has seen as much quiet and order, and more honesty in paying just debts, than where legal restraints operated in all their force. The turpitude of vice and the majesty of virtue, were as apparent as in older settlements. Industry, in laboring or hunting, bravery in war, candor, honesty and hospitality were rewarded with the confidence and honor of the people. Regulating parties would exist, and thieves, rogues and counterfeiters were sure to receive a striped jacket, "worked nineteen to the dozen;" and by this mode of operation, induced to "clear out:" but truth, uprightness, honesty and sincerity are always respected. Many of the frontier class are *illiterate*, but they are, by no means, *ignorant*. They are a shrewd, observing, thinking people. They may not have learned the black marks in books, but they have studied *men* and *things*, and have a quick insight into human nature. They are not inattentive to religion, though their opportunities of religious instruction are few compared with old countries. They have prejudices and fears about many of the organized benevolent societies of the present age, yet there are no people more readily disposed to attend religious meetings, and whose hearts are more readily affected with the gospel, than the backwoods people; and as large a proportion are orderly professors of religion as in any part of the Union. Ministers of the gospel and missionaries, who can suit themselves to the circumstances and habits of frontier people; who, like Paul, can "become all things to all men," find pleasant and interesting fields of labor on all our frontiers. But let such persons show fastidiousness, affect superior intelligence and virtue, catechise the people for their plainness and simplicity of manners, and draw invidious comparisons, and they are sure to be "used up," or left without hearers, to deplore the "dark clouds" of ignorance and prejudice in the West.

Hunters and Trappers. Entirely beyond the boundaries of civilization, are many hundreds of an unique class, distinguished by the terms hunters and trappers. They are engaged in hunting buffalo, and other wild game, and trapping for beaver. They are found upon the vast prairies of the West and north-west; in all the defiles, and along the streams of the Rocky mountains, and in various parts of the Oregon Territory, to the peninsula of California. They are an enterprising and erratic race, from almost every State, and are usually in the employ of persons of capital and enterprise, and who are concerned in the fur and peltry business. Expeditions for one, two, or three years, are fitted out from St. Louis, or some commercial point, consisting of companies, who ascend the rivers to the regions of fur. The hunters and trappers receive a proportion of the profits of the expedition. Some become so enamored with this wandering and exposed life, as to lose all desire of returning to the abodes of civilization, and remain for the rest of their lives in the American deserts. There are individuals, who are graduates of colleges, and who once stood high in the circles of refinement and taste, that have passed more than twenty years amongst the roaming tribes of the Rocky mountains, or on the western slope, till they have apparently lost all feelings towards civilized life. They have afforded an interesting but melancholy example of the tendencies of human nature towards the degraded state of savages. The improvement of the species is a slow and laborious process;—the deterioration is rapid, and requires only to be divested of restraint, and left to its own unaided tendencies. Many others have returned to the habits of civilization, and some with fortunes made from the woods and prairies.

Boatmen. These are the fresh-water sailors of the West, with much of the light-hearted, reckless character of the sons of the ocean, including peculiar shades of their own. Before the introduction of steam-boats on the western waters, its immense commerce was carried on by means of *keel-boats* and *barges*. The former is much in the shape of a canal-boat, long, slim-built, sharp at each end, and propelled by setting-poles, and the cordelle or long rope. The barge is longer, and has a bow and stern. Both are calculated to ascend streams, but by a very slow process. Each boat would require from ten to thirty hands, according to its size. A number of these boats frequently sailed in company. The boatmen were proverbially lawless, at every town and landing, and indulged without restraint in every species of dissipation, debauchery and excess. But this race has become reformed, or nearly extinct;—yes, reformed, by the mighty power of steam. A steam-boat, with half the crew of a barge or keel, will carry ten times the burden, and perform six or eight trips in the time it took a keel-boat to make one voyage. Thousands of flat boats, or "broad horns," as they are called, pass *down* the rivers, with the produce of the country, which are managed by the farmers of the West, but never return *up* stream. They are sold for lumber, and the owners, after disposing of the cargo, return by steam. The number of boatmen on the western waters is not only greatly reduced, but those that remain are fast losing their original character.

CHAPTER V.

PUBLIC LANDS.

System of Surveys—Meridian and Base Lines—Townships—Diagram of a Township, surveyed into Sections—Land Districts and Offices—Preëmption Rights—Military Bounty Lands—Taxes—Valuable Tracts of Country unsettled.

In all the new States and Territories, the lands which are owned by the general government, are surveyed and sold under one general system. Several offices, each under the direction of a surveyor-general, have been established by acts of Congress, and districts, embracing one or more States, assigned them. The office for the surveys of all public lands in Ohio, Indiana, Michigan and the Wisconsin country, is located at Cincinnati. The one including the States of Illinois, Missouri and Arkansas, is at St. Louis. Deputy-surveyors are employed to do the work at a stipulated rate per mile, generally from three to four dollars, who employ chain-bearers, an axe and flag man, and a camp-keeper. They are exposed to great fatigue and hardships, spending two or three months at a time in the woods and prairies, with slight, movable camps for shelter.

In the surveys, *meridian* lines are first established, running north from the mouth of some noted river: these are intersected with *base* lines.

There are five principal meridians in the land-surveys in the West.

The *first principal meridian* is a line due north from the mouth of the Miami.

The *second principal meridian* is a line due north from the mouth of Little Blue river, in Indiana.

The *third principal meridian* is a line due north from the mouth of the Ohio.

The *fourth principal meridian* is a line due north from the mouth of the Illinois.

The *fifth principal meridian* is a line due north from the mouth of the Arkansas. Another meridian is used for Michigan, which passes through the central part of the State. Its base line extends from about the middle lake St. Clair, across the State, west, to lake Michigan. Each of these meridians has its own base line.

The surveys connected with the third and fourth meridians, and a small portion of the second, embrace the State of Illinois.

The base line for both the second and third principal meridians commences at Diamond Island, in Ohio, opposite Indiana, and runs due west, till it strikes the Mississsippi, a few miles below St. Louis.

All the *townships* in Illinois, south and east of the Illinois river, are numbered from this base line, either north or south.

The third principal meridian terminates with the northern boundary of the State.

The fourth principal meridian commences in the centre of the channel, and at the mouth of the Illinois river, but immediately crosses to the *east* shore, and passes up on that side (and at one place nearly fourteen miles distant), to a point in the channel of the river, seventy-two miles from its mouth. Here its base line commences, and extends across the peninsula to the Mississippi, a short distance above Quincy. The fourth principal meridian is continued northward through the military tract, and across Rock river, to a curve in the Mississippi, at the upper rapids, in township eighteen north, and about twelve or fifteen miles above Rock island. It here crosses and passes up the *west* side of the Mississippi river, fifty-three miles, and recrosses into Illinois, and passes through the town of Galena, to the northern boundary of the State. It is thence continued to the Wisconsin river, and made the principal meridian for the surveys of the Territory, while the northern boundary line of the State is constituted its base line for that region.

Having formed a principal meridian, with its corresponding base line, for a district of country, the next operation of the surveyor is to divide this into tracts of six miles square, called *townships*.

In numbering the townships *east* or *west* from a principal meridian, they are called *ranges*, meaning a range of townships; but in numbering *north* or *south* from a base line, they are called *townships*. Thus a tract of land is said to be situated in township four north, in range three east, from the third principal meridian; or as the case may be.

Townships are subdivided into square miles, or tracts of six hundred and forty acres each, called *sections*. If near timber, trees are marked and numbered with the section, township and range, near each sectional corner. If in a large prairie, a mound is raised to designate the corner, and a billet of charred wood buried, if no rock is near. Sections are divided into halves, by a line north and south, and into quarters, by a transverse line. In sales under certain conditions, quarters are sold in equal subdivisions of forty acres each, at $1 25 per acre. Any person, whether a native-born citizen, or a foreigner, may purchase forty acres of the richest soil, and receive an indisputable title, for fifty dollars.

Ranges are townships, counted either east or west from meridians.

Townships are counted either north or south from their respective base lines.

Fractions are parts of quarter sections intersected by streams, or confirmed claims.

The parts of townships, sections, quarters, &c., made at the lines of either townships or meridians, are called *excesses* or *deficiences*.

Sections, or square miles, are numbered,

beginning in the north-east corner of the township, progressively west, to the range line, and then progressively east, to the range line, alternately, terminating at the south-east corner of the township, from one to thirty-six, as in the following diagram:

6	5	4	3	2	1
7	8	9	10	11	12
18	17	16*	15	14	13
19	20	21	22	23	24
30	29	28	27	26	25
31	32	33	34	35	36

I have been thus particular in this account of the surveys of public lands, to exhibit the simplicity of a system, that, to strangers, unacquainted with the method of numbering the sections, and the various subdivisions, appears perplexing and confused.

All the lands of Congress, owned in Ohio, have been surveyed, and, with the exception of some Indian reservations, have been brought into market. In Indiana, all the lands purchased of the Indians have been surveyed, and, with the exception of about ninety townships and fractional townships, have been offered for sale. These, amounting to about two millions of acres, will be

* Appropriated for schools in the township.

offered for sale the present year. In Michigan, nearly all the ceded lands have been surveyed and brought into market. The unsurveyed portion is situated in the neighborhood of Saginau bay; a part of which may be ready for market within the current year.

In Wisconsin Territory, west of lake Michigan, all the lands in the Wisconsin district, which lie between the State of Illinois and the Wisconsin river, have been surveyed; and, in addition to the lands already offered for sale in the Green Bay district, about sixty-five townships and fractional townships have been surveyed, and are ready for market. The surveys of the whole country west of lake Michigan, and south of the Wisconsin river, in Illinois and Wisconsin Territory, will soon be surveyed and in market. Here are many millions of the finest lands on earth, lying along the Des Plaines, Fox and Rock rivers, and their tributaries, well watered, rich soil, a healthy atmosphere, and facilities to market. A temporary scarcity of timber in some parts of this region will retard settlements, for a time; but this difficulty will be obviated, by the rapidity with which prairie land turns to a timbered region, wherever, by contiguous settlements, the wild grass becomes subdued, and by the discovery of coal-beds. Much of it is a mineral region. In Illinois; the surveys are now completed in the Danville district, and in the southern part of the Chicago district. They are nearly completed along Rock river and the Mississippi.

The unsurveyed portion is along Fox river, Des Plaines, and the shore of lake Michigan, in the north-eastern part of the State. Emigrants, however, do not wait for surveys and sales. They are settling over this fine portion of the State, in anticipation of purchases. In Missouri, besides the former surveys, the exterior lines of one hundred and thirty-eight townships, and the subdivision into sections and quarters, of thirty townships, in the northern part of the State, and contracts for running the exterior lines of one hundred and eighty-nine townships, on the waters of the Osage and Grand rivers, have been made. A large portion of this State is now surveyed, and in market. Surveys are progressing in Arkansas, and large bodies of land are proclaimed for sale, in that district.

I have no data before me that will enable me to show, definitely, the amount of public lands now remaining unsold, in each land-office district. In another place, I have already given an estimate of the amount of public lands, within the organized States and Territories, remaining unsold, compared with the amount sold in past years.

The following tables exhibit the number of acres sold in the districts embraced more immediately within the range of this Guide, for 1834, and the three first quarters of 1835, with the name of each district in each State. It is constructed from the report of the commissioner of the General Land Office to the Treasury Department, December 5th, 1835.

The sales of the last quarter of 1835, in Illinois, and probably in the other States, greatly exceeded either of the other quarters, and which will be exhibited in the annual report of the commissioner, in December, 1836.

Statement of the amount of Public Lands sold at the several Land-Offices in Ohio, Indiana, Illinois, Michigan, Wisconsin, Missouri and Arkansas, in 1834.

District Land Offices.	Acres and hundredths.
OHIO.	
Marietta, ...	11,999.52
Zanesville, ...	33,877.23
Steubenville, . . .	4,349.19
Chilicothe, ...	21,309.32
Cincinnati, ...	27,369.52
Wooster, ...	9,448.77
Wapaghkonetta,	125,417.13
Bucyrus, ...	245,078.56
Total for State,	478,847.24
INDIANA.	
Jeffersonville, . .	67,826.11
Vincennes, ...	56,765.80
Indianapolis, . . .	204,526.63
Crawfordsville, .	161,477.87
Fort Wayne, . . .	96,350.30
La Porte, ...	86,709.73
Total for State,	673,656.44
ILLINOIS.	
Shawneetown, .	6,904.24
Kaskaskia, ...	15,196.52
Edwardsville, . .	124,302.19
Vandalia, ...	20,207.61
Palestine, ...	22,135.69
Springfield, ...	66,804.25
Danville, ...	62,331.38
Quincy, ...	36,131.59
Total for State,	354,013.47
MICHIGAN.	
Detroit, ...	136,410.69
Monroe, ...	233,768.30
White Pigeon Prarie and Bronson,	128,244.47
Total for Ter.,	498,423.46
WISCONSIN.	
Mineral Point, . . .	14,336.67
MISSOURI.	
St. Louis, ...	43,634.68
Fayette, ...	71,049.74
Palmyra, ...	76,241.35
Jackson, ...	18,882.11
Lexington, ...	43,983.80
Total for State,	253,791.70
ARKANSAS.	
Batesville, ...	8,051.31
Little Rock, ...	25,799.74
Washington, ...	65,145.88
Fayetteville, ...	24,514.94
Helena, ...	26,244.59
Total for Ter.,	149,756.46

Statement of the amount of Public Lands sold at the several Land-Offices in Ohio, Indiana, Illinois, Michigan, Wisconsin, Missouri and Arkansas, from Jan. 1st, to Sept. 30th, 1835, including nine months.

District Land Offices.	Acres and hundredths.
OHIO.	
Marietta, ...	11,012.98
Zanesville, ...	42,978.36
Steubenville, . . .	3,649.29
Chilicothe, ...	12,586.87
Cincinnati, ...	20,105.76
Wooster, ...	5,157.68
Wapaghkonetta and Lima,	103,020.23
Bucyrus, ...	154,706.63
Total for State,	353,217.80
INDIANA.	
Jeffersonville, . .	44,634.81
Vincennes, ...	70,903.62
Indianapolis, . . .	158,786.68
Crawfordsville, .	108,055.22
Fort Wayne, . . .	148,864.28
La Porte, ...	227,702.35
Total for State,	758,946.96
ILLINOIS.	
Shawneetown, .	5,754.08
Kaskaskia, ...	13,814.38
Edwardsville, . .	123,638.07
Vandalia, ...	16,253.46
Palestine, ...	14,088.01
Springfield, ...	316,966.70
Danville, ...	94,491.35
Quincy, ...	*40,274.58
Galena, ...	†262,152.73
Chicago, ...	333,405.40
Total for State,	1,220,838.76
MICHIGAN.	
Detroit, ...	213,763.57
Bronson, ...	400,722.48
Monroe, ...	446,631.61
Total,	1,061,127.66
WISCONSIN.	
Mineral Point, . . .	67,052.55
Green Bay, ...	68,365.53
Total for Ter.,	135,418.08
MISSOURI.	
St. Louis, ...	32,914.57
Fayette, ...	55,839.58
Palmyra, ...	101,018.00
Jackson, ...	28,995.19
Lexington, ...	42,801.45
Springfield, ...	320.00
Total for State,	261,888.79
ARKANSAS.	
Batesville, ...	2,021.22
Little Rock, ...	22,291.92
Washington, ...	43,360.81
Fayetteville, ...	8,723.72
Helena, ...	312,169.09
Total for State,	388,566.76

* Returns only to May 31. † Returns only to July 31.
Since those periods, the sales at these offices have been immense.

The reader will perceive that the sales of the first three quarters of 1835 almost doubled those of the whole year of 1834. The inquiry was often made of the writer, while traveling in the Atlantic States in the summer of 1835, whether there was still opportunity for emigrants to purchase public lands in Indiana, Illinois, &c., where land-offices had been opened for the sale of lands many years. He found, almost every where, wrong notions prevailing. The people were not aware of the immense extent of the public domain now in market, and ready to be sold at *one dollar and twenty-five cents per acre*, and even in as small tracts as forty acres. Take, for example, the Edwardsville district, in which the writer resides. It extends south to the base line, east to the third principal meridian, north to the line that separates townships 13 and 14 north, and west to the Illinois and Mississippi rivers, and embraces all the counties of Madison, Clinton, Bond, Montgomery, Macoupen, and Greene, a tier of townships on the south side of Morgan and Sangamon, five and a half townships from Fayette, and about half of St. Clair county. The lands for a part of this district have been in market for eighteen or twenty years. It contains some of the oldest American settlements in the State, and has also a number of confirmed claims never offered for sale. And yet the receiver of this office informed me, in November last, that he had just made returns of all the lands sold in this district, and they amounted to just *one third* of the whole quantity. Every man, therefore, may take it for granted that there will be land enough in market in all the new States, for his use, during the present generation. These are facts that should be known to all classes. The mania of land speculation and of monopolists would soon subside, were those concerned to sit down coolly, and, after ascertaining the amount of public lands now in market, with the vast additional quantity that must soon come into market, use a few figures in common arithmetic, with the probable amount of emigration, and ascertain the probable extent of the demand for this article at any future period.

The following information is necessary for those who are not acquainted with our land system.

In each land-office there are a register and receiver, appointed by the president and senate for the term of four years, and paid by the government.

After being surveyed, the land, by proclamation of the president, is offered for sale at public auction by half quarter sections, or tracts of eighty acres. If no one bids for it at one dollar and twenty-five cents per acre, or more, it is subject to private entry at any time after, upon payment of $1 25 cents per acre at the time of entry. *No credit, in any case, is allowed.*

In many cases, Congress, by special statute, has granted to actual settlers, preëmption rights, where settlements and improvements have been made on public lands previous to public sale.

Preëmption rights confer the privilege only of purchasing the tract containing improvements at one dollar and twenty-five cents per acre, by the possessor, without the risk of a public sale.

In Illinois, and several other Western States, all lands purchased of the general government, are exempted from taxation for five years after purchase.

Military Bounty Lands. These lands were surveyed and appropriated as bounties to the soldiers in the war with Great Britain in 1812—'15, to encourage enlistments. The selections were made in Illinois, Missouri, and Arkansas. The bounty lands of Illinois lie between the Illinois and Mississippi rivers, in the counties of Calhoun, Pike, Adams, Schuyler, Macdonough, Warren, Mercer, Knox, Henry, Fulton, Peoria, and Putman. Out of 5,000,000 of acres, 3,500,000 were selected, including about three fifths of this tract. The remainder is disposed of in the manner of other public lands. The disposition of this fine country for military bounties has much retarded its settlement. It was a short-sighted and mistaken policy of government that dictated this measure. Most of the titles have long since departed from the soldiers for whose benefit the donations were made. Many thousand quarter sections have been sold for taxes by the State, have fallen into the hands of monopolists, and are now past redemption. The bounty lands in Missouri lie on the waters of Chariton and Grand rivers, north side of the Missouri river, and in the counties of Chariton, Randolph, Carroll, and Ray, and include half a million of acres. The tract is generally fertile, undulating, a mixture of timber and prairie, but not as well watered as desirable. With the bounty lands of Arkansas I am not well acquainted. Their general character is good, and some tracts are rich cotton lands.

Taxes. Lands bought of the United States government are exempted from taxation for five years after sale. All other lands owned by non-residents, equally with those of residents, are subject to taxation annually, either for state, or county purposes, or both. The mode and amount varies in each State. If not paid when due, costs are added, the lands sold, subject to redemption within a limited period,—generally two years. Every non-resident landholder should employ an agent within the State where his land lies, to look after it, and pay his taxes, if he would not suffer the loss of his land.

CHAPTER VI.

ABORIGINES.

Conjecture respecting their former Numbers and Condition—Present Number and State—Indian Territory appropriated as their permanent Residence—Plan and Operations of the United States Government—Missionary Efforts and Stations—Monuments and Antiquities.

The idea is entertained, that the Valley of the Mississippi was once densely populated by aborigines,—that here were extensive nations,—that the bones of many millions lie mouldering under our feet. It has become a common theory, that, previous to the settlement of the country by people of European descent, there were *two* successive races of men, quite distinct from each other;—that the first race, by some singular fatality, became exterminated, leaving no traditionary account of their existence; and the second race, the ancestors of the existing race of Indians, are supposed to have been once far more numerous than the present white population of the Valley.

Some parts of Mexico and South America

were found to be populous upon the first visits of the Spaniards, but I do not find satisfactory evidence that the population was ever dense in any part of the territory that now constitutes our Republic. Mr. Atwater supposes, from the mounds in Ohio, the Indian population far exceeded seven hundred thousand, at one time, in that district. Mr. Flint says, "If we can infer nothing else from the mounds, we can clearly infer, that this country once had its millions." Hence, a principal argument assigned for the populousness of this country is, the millions buried in these tumuli, the bones of which, in a tolerable state of preservation, are supposed to be exhibited upon excavation. The writer has witnessed the opening of many of these mounds, and has seen the fragments of an occasional skeleton, found *near the surface*. Without stopping here to enter upon a disquisition on the hypothesis assumed,—that these mounds, as they are termed, are as much the results of natural causes, as any other prominences on the surface of the globe,—I will only remark, that it is a fact well known to frontier men, that the Indians have been in the habit of burying their dead in these ridges and hillocks, and that, in our light, spongy soil, the skeleton decays surprisingly fast. This is not the place to exhibit the necessary data, that have led to the conviction, that not a human skeleton now exists in all the Western Valley (excepting in nitrous caves), that was deposited in the earth before the discovery of the new world by Columbus.

The opinion, that this Valley was once densely populous, is sustained, from the supposed military works, distributed through the West. This subject, as well as that of mounds, wants reëxamination. Probably, half a dozen enclosures, in a rude form, might have been used for military defence. The capabilities of the country to sustain a dense population, has been used to support the position, that it must have been once densely populated. This argues nothing without vestiges of agriculture and the arts. With the exception of a few small patches round the Indian villages, for corn and pulse, the whole land was an unbroken wilderness. Strangers to the subject have imagined that our western prairies must once have been subdued by the hand of culitvation, because denuded of timber. Those who have long lived on them, have the evidences of observation and their senses to guide them. They know, that the earth will not produce timber while the surface is covered with a firm grassy sward, and that timber will spring up, as soon as this obstruction is removed.

To all these theories, of the former density of the aboriginal population of the Valley, I oppose, first, the fact that but a scattered and erratic population was found here, on the arrival of the Europeans; that the people were rude savages, subsisting chiefly by hunting, and that no savage people ever became populous; that, from time immemorial, the different tribes had been continually at war with each other; that, but a few years before the French explored it, the Iroquois, or Five Nations, conquered all the country to the Mississippi, which they could not have done had it been populous; and that Kentucky, one of the finest portions of the Valley, was not inhabited by any people, but was the common hunting and fighting grounds of both the northern and southern Indians, and hence called by them *Kentuckee*, or the "bloody ground."*

That the Indian character has deteriorated, and the numbers of each tribe greatly lessened by contact with Europeans and their descendants, is not questioned; but many of the descriptions of the comforts and happiness of savage life and manners, before their country was possessed by the latter, are the exaggerated and glowing descriptions of poetic fancy. Evidence enough can be had to show that they were degraded and wretched, engaged in petty exterminating wars with each other, oftentimes in a state of starvation, and leading a roving, indolent, and miserable existence. Their government was anarchy. Properly speaking, civil government had never existed amongst them. They had no executive, or

* See Pownal's Administration of the British Colonies; Colden's History of the Five Nations; New York Historical Collections, vol. II; Charlevoix Histoire de la Nouvelle France; Hon. De Witt Clinton's Discourse before the New York Historical Society, 1811; Discovery of the Mississippi river, by Father Lewis Hennepin; M. Tonti's Account of M. De La Salle's Expedition; La Harpe's Journal, &c.

judiciary power, and their legislation was the result of their councils, held by aged and experienced men. It had no stronger claim upon the obedience of the people than advice.

In Mexico, civilization had made progress, and there were populous towns and cities, and edifices for religious and other purposes. With the exception of some very rude structures, the ruins of which yet remain, and which, upon too slight grounds, have been mistaken for military works, nothing is left as marks of the enterprise of the feeble bands of Indians of this Valley. Their implements, utensils, weapons of war, and water-craft, were of the most rude and simple construction, and yet prepared with great labor. Those who have written upon Indian manners, without personal and long acquaintance with their circumstances, have made extravagant blunders. The historian of America, Dr. Robertson, seems to suppose that the Indians cut down large trees and dug out canoes, with stone hatchets; and that they cleared the timber from their small fields, by the same tedious process. Their stone axes or hatchets were never used for *cutting*, but only for splitting and pounding. They burned down and hollowed out trees, by fire, for canoes; and never chopped off the timber, but only deadened it, in clearing land. The condition of depraved man, unimproved by habits of civilization, and unblest with the influences and consolations of the gospel, is pitiable in the extreme. Such was the character and condition of the "red skin," before his land was visited by the "pale faces." I have often seen the aboriginal man in all his primeval wildness, when he first came in contact with the evils and benefits of civilization,—have admired his noble form and lofty bearing,—listened to his untutored and yet powerful eloquence, and yet have found in him the same humbling and melancholy proofs of his wretchedness and want, as is found in the remnants on our borders.

The introduction of ardent spirits, and of several diseases, are the evils furnished the Indian race, by contact with the whites, while in other respects their condition has been improved.

From the second number of the *Annual Register of Indian Affairs, within the Indian (or Western) Territory*, just published by the Rev. Isaac McCoy, the following particulars have been chiefly gleaned.

Mr. McCoy has been devoted to the work of Indian reform for almost twenty years; first in Indiana, then in Michigan, and latterly in the Indian Territory, west of Missouri and Arkansas. He is not only intimately acquainted with the peculiar circumstances of this unfortunate race, and with the country selected as their future residence by the government, but is ardently and laboriously engaged for their welfare.

Indian Territory.

The Indian Territory lies west, and immediately adjacent to Missouri and Arkansas. It is about six hundred miles long, from north to south, extending from the Missouri river to the Red river, and running westwardly as far as the country is habitable, which is estimated to be about two hundred miles. The almost destitution of timber, with extensive deserts renders most of the country, from this Territory to the Rocky mountains, uninhabitable. The dreams indulged by many, that the wave of white population is to move onward, without any resisting barrier, till it reaches these mountains, and even overleap them, to the Pacific ocean, will never be realized. Providence has thrown a desert of several hundred miles in extent, as an opposing barrier.

As very contradictory accounts have gone abroad, prejudicial to the character of the country selected for the Indians, it becomes necessary to describe it with some particularity. The following, from Mr. McCoy (if it needed any additional support to its correctness), is corroborated by the statements of many disinterested persons.

"There is a striking similarity between all parts of this Territory. In its general character, it is high and undulating, rather level than hilly; though small portions partly deserve the latter appellation. The soil is generally very fertile. It is thought that in no part of the world, so extensive a region of rich soil has been discovered, as in this, of which the Indian Territory is a central position. It is watered by numerous rivers, creeks and rivulets. Its waters pass through it eastwardly, none of which are favorable to navigation. There is less marshy and stagnant water in it than is usual in the western country. The atmosphere is salubrious, and the climate precisely such as is desirable, being about the same as that inhabited by the Indians on the east of the Mississippi. It contains much mineral coal and salt water, some lead, and some iron ore. Timber is scarce, and this is a serious defect, but one which time will remedy, as has been demonstrated by the growth of timber in prairie countries which have been settled, where the grazing of stock, by diminishing the quantity of grass, renders the annual fires less destructive to the growth of wood. The prairie (i. e., land destitute of wood) is covered with grass, much of which is of suitable length for the scythe."

The *Choctaws*, *Creeks*, *Cherokees*, *Osages*, *Kausaus*, and *Delawares*, are entitled to lands westward of this Territory, for hunting-grounds; some to the western boundary of the United States, others to the Rocky mountains.

Mr. McCoy estimates the number of inhabitants of this Territory at 47,733, viz.:

Indigenous Tribes.

Osage, about	5,510
Kausau, "	1,684
Ottoe and Missourias,	1,600
O'Mahaus,	1,400
Pawnees, (four tribes),	10,000
Puncahs, about	800
Quapaws, "	450
	21,444

Emigrant Tribes.

Choctaw, about	15,000
Cherokee, "	4,000
Creek, "	3,600
Seneca, Shawanoe of Neosho,	462
Wea, about	225
Piankeshau,	119
Peoria and Kaskaskias,	135
Ottawa,	81
Shawanoe, of Kausau river,	764
Delaware,	856
Kickapoo,	603
Putawatomie,	444
Emigrants,	26,289
Indigenous,	21,444
Total,	47,733

The estimate of the Choctaws includes about four hundred negro slaves; that of the Cherokees, five hundred; and that of the Creeks about four hundred and fifty slaves.

Choctaws. Their country adjoins Red river and the Province of Texas, on the south, Arkansas on the east, and extends north to the Arkansas and Canadian rivers; being one hundred and fifty miles from north to south, and two hundred miles from east to west. Here are numerous salt springs. For civil purposes, their country is divided into three districts.

Cherokees. The boundaries of their country commences on the Arkansas river, opposite the western boundary of Arkansas; thence northwardly along the line of Missouri, eight miles to Seneca river; thence west to the Neosho river; thence up said river to the Osage lands; thence west indefinitely, as far as habitable; thence south to the Creek lands, and along the eastern line of the Creeks to a point forty-three miles west of Arkansas, and twenty-five miles north of Arkansas river; thence to the Verdigris river, and down Arkansas river, to the mouth of the Neosho; thence southwardly to the junction of the North Fork and Canadian rivers; and thence down the Canadian and Arkansas rivers to the place of beginning. The treaty of 1828 secures to this tribe 7,000,000 of acres, and adds land westward, for hunting-grounds, as far as the United States boundaries extend.

The *Creeks* or *Muscogees*, occupy the country west of Arkansas, that lies between the lands of the Choctaws and Cherokees.

The *Senecas* join the State of Missouri on the east, with the Cherokees south, the Neosho river west, and possess 127,500 acres.

The *Osage* (a French corruption of *Wos-sosh-e* their proper name) have their country north of the western portion of the Cherokee lands, commencing twenty-five miles west of the State of Missouri, with a width of fifty miles, and extending indefinitely west. About half the tribe are in the Cherokee country.

The *Quapaws* were originally connected with the Osages. They have migrated from the Lower Arkansas, and have their lands adjoining the State of Missouri, immediately north of the Senecas.

The *Putawatomies* are on the north-eastern side of the Missouri river, but they are not satisfied, and the question of their locality is not fully settled. Four hundred and forty-four Putawatomies are mingled with the Kickapoos, on the south-west side of the Missouri river.

The *Weas*, *Piankeshaus*, *Peorias* and *Kaskaskias* are remnants of the great western confederacy, of which the Miamies were the most prominent branch. These and other tribes, constituted the Illini, Oillinois, or Illinois nation, that once possessed the country now included in the great States of Indiana, Illinois, &c. Their lands lie west of the State of Missouri, and south-west of the Missouri river.

The *Delawares* occupy a portion of the country in the forks of the Kausau (or, as written by the French, Kausas) river. They are the remnants of another great confederacy, the *Lenni-Lenopi*, as denominated by themselves.

The lands of the *Kickapoos* lie north of the Delawares, and along the Missouri, including 768,000 acres.

The *Ottoes* occupy a tract of country between the Missouri and Platte rivers, but their land is said to extend south and below the Platte.

The country of the *O'Mahaus* has the Platte river on the south, and the Missouri north-east.

The country of the *Pawnees* lies to the westward of the Ottoes and O'Mahaus. The boundaries are not defined.

The *Puncahs* are a small tribe that originated from the Pawnees, and live in the northern extremity of the country spoken of as the Indian Territory.

Present Condition. The Choctaws, Cherokees and Creeks are more advanced in civilized habits than any other tribes. They have organized local governments of their own, have enacted some wholesome laws, live in comfortable houses, raise horses, cattle, sheep and swine, cultivate the ground, have good fences, dress like Americans, and manufacture much of their own clothing. They have schools and religious privileges, by missionary efforts, to a limited extent. The Cherokees have a written language, perfect in its form, the invention of Mr. Guess, a full-blooded Indian. The Senecas, Delawares, and Shawanoes, also, are partially civilized, and live with considerable comfort from the produce of their fields and stock. The Putawatomies, Weas, Piankeshaus, Peorias, Kaskaskias, Ottawas, and Kickapoos, have partially adopted civilized customs. Some live in comfortable log cabins, fence and cultivate the ground, and have a supply of stock: others live in bark huts, and are wretched. The Osages, or Wos-sosh-es, Quapaws, Kausaus, Ottoes, O'-Mahaus, Pawnees and Puncahs have made much less improvement in their mode of living. A few have adopted civilized habits, and are rising in the scale of social and individual comforts, but the larger portion are yet *Indians*.

Mr. McCoy estimates the whole number of aborigines in North America, including those of Mexico, at 1,800,000; of which, 10,000 are so far improved as to be classed with civilized men, and amongst whom there are as many pious Christians, as amongst the same amount of population in the United States. In addition to these, he estimates that there may be about 60,000 more, "which may have made advances toward civilization, some more and some less."

For some years past, the policy of the government of the United States has been directed to the project of removing all the Indians from the country organized into States and Territories, and placing them sufficiently contiguous to be easily governed, and yet removed from direct contact and future interruption from the white population. This project was recommended in the period of Mr. Monroe's administration; was further considered, and some progress made, under that of Mr. Adams; but has been carried into more successful execution within the last five years. It is much to be regretted that this project was not commenced earlier. The residence of small bands of Indians, with their own feeble and imperfect government, carried on within any organized State or Territory, is ruinous. Those who argue that *because* of the removal of the Indians from within the jurisdiction of the States, or an organized Territory, *therefore* they will be driven back from the country in which it is now proposed to place them, evince but a very partial and imperfect view of the subject. The present operation of government is an experiment; and it is one that ought to receive a fair and full trial. If it does not succeed, I know not of any governmental regulation that can result, with success, to the prosperity of the Indians. The project is to secure to each tribe, by patent, the lands allotted them; to form them into a territorial government, with some features of the representative principle; to have their whole country under the supervision of our government, as their guardian, for their benefit; to allow no white men to pass the lines and intermix with the Indians, except those who are licensed by due authority; to aid them in adopting civilized habits, provide for them schools and other means of improving their condition, and, through the agency of missionary societies, to instruct them in the principles of the gospel of Christ.

Missionary Efforts and Stations. These are conducted by the American Board of Commissioners for Foreign Missions; the Baptist

Board of Foreign Missions; the Methodist Episcopal Missionary Society; the Western Foreign Missionary Society; and the Cumberland Presbyterians. Stations have been formed, and schools established, with most of these tribes. About 2500 are members of Christian churches, of different denominations. The particulars of these operations are to be found in the reports of the respective societies, and the various religious periodicals.

Of other tribes within the Valley of the Mississippi, and not yet within the Indian Territory, the following estimate is sufficiently near the truth for practical purposes:—

Indians from New York, about Green Bay,..	725
Wyandots, in Ohio and Michigan,........	623
Miamies,.........................	1,200
Winnebagoes,.....................	4,591
Chippeways, or O'Jibbeways,..........	6,793
Ottawas and Chippeways, of lake Michigan,..	5,300
Chippeways, Ottawas and Putawatomies,....	8,000
Putawatomies,.....................	1,400
Menominees,......................	4,200

They are all east of the Mississippi, and chiefly found on the reservations in Ohio, Indiana and Michigan, and in the country between the Wisconsin river and lake Superior. Those tribes west of the Mississippi river, and along the region of the Upper Missouri river, are as follows:—

Sioux,...........................	27,500
Ioways,..........................	1,200
Sauks, of Missouri,................	500

Sauks and Foxes,	6,400
Assinaboines,	8,000
Crees,	3,000
Gros Ventres,	3,000
Aurekaras,	3,000
Cheyennes,	2,000
Mandans,	1,500
Black Feet,	30,000
Camanches,	7,000
Minatarees,	1,500
Crows,	4,500
Arrepahas and Kiawas,	1,400
Caddoes,	800
Snake, and other tribes, within the Rocky mountains,	20,000
West of the Rocky mountains,	80,000

The Camanches, Arrepahas, Kiawas and Caddoes roam over the great plains, toward the sources of the Arkansas and Red rivers, and through the northern parts of Texas. The Black Feet are toward the heads of the Missouri.

Monuments and Antiquities. Before dismissing the subject of the aborigines, I shall touch very briefly on the monuments and antiquities of the West; with strong convictions that there has been much exaggeration on this subject. I have already intimated that the mounds of the West are natural formations, but I have not room for the circumstances and facts that go to sustain this theory. The number of objects considered as antiquities is greatly exaggerated. The imaginations of men have done much. The number of mounds on the American bottom, in Illinois, adjacent to Cahokia creek, is stated by Mr. Flint at two hundred. The writer has counted all the elevations of surface for the extent of nine miles, and they amount to seventy-two. One of these, Monk hill, is much too large, and three fourths of the rest are quite too small for human labor. The pigmy graves on the Merrimeek, (Missouri,) in Tennessee, and other places, upon closer inspection, have been found to contain decayed skeletons of the ordinary size, but buried with the leg and thigh bones in contact. The *giant* skeletons sometimes found, are the bones of buffalo.

It is much easier for waggish laborers to deposit old horse-shoes, and other iron articles, where they are at work, for the special pleasure of digging them up for credulous antiquarians, than to find proofs of the existence of the horses that wore them!

There may, or may not be, monuments and antiquities that belong to a race of men of prior existence to the present race of Indians. All that the writer urges is, that this subject may not be considered as settled; that due allowance may be made for the extreme credulity of some, and the want of personal observation and examination of other writers on this subject. Gross errors have been committed, and exaggerations of very trivial circumstances have been made.

The antiquities belonging to the Indian race are neither numerous or interesting, unless we except the remains of rude edifices and enclosures, the walls of which are almost invariably embankments of earth. They are rude axes and knives of stone, bottles and vessels of potter's ware, arrow and spear heads, rude ornaments, &c.

Roman, French, Italian, German and English coins and medals, with inscriptions, have been found; most unquestionably brought by Europeans; probably by the Jesuits, and other orders, who were amongst the first explorers of the West, and who had their religious houses here, more than a century past.

Copper and silver ornaments have been discovered in the mounds that have been opened. The calumet, or large stone pipe, is often found in Indian graves. Three facts deserve to be regarded by those who examine mounds and Indian cemeteries. 1. That the Indians have been accustomed to bury their dead in these mounds. 2. That they were accustomed to place various ornaments, utensils, weapons, and other articles of value, the property of the deceased, in these graves, especially if a chieftain, or man of note. A third fact, known to our frontier people, is the custom of several Indian tribes wrapping their dead in strips of bark, or encasing them with the halves of a hollow log, and placing them in the forks of trees. This was the case especially when their deaths occurred while on hunting or war parties. At stated seasons, these relics were collected, with much solemnity, brought to the common sepulchre of the tribe, and deposited with their ancestors. This accounts for the confused manner in which the bones are often found in mounds and Indian grave-yards. Human skeletons, or rather mummies, have been discovered in the nitrous caves of Kentucky. The huge bones of the mammoth and other enormous animals, have been exhumed at the Bigbone licks, in Kentucky, and in other places.

CHAPTER VII.

WESTERN PENNSYLVANIA.

Face of the Country—Soil, Agriculture and internal Improvements—Chief Towns—Pittsburgh—Coal.

The portion of Pennsylvania lying west of the Alleghany ridge, contains the counties of Washington, Greene, Fayette, Westmoreland, Alleghany, Beaver, Butler, Armstrong, Mercer, Venango, Crawford, Erie, Warren, McKean, Jefferson, Indiana, Somerset, and a part of Cambria.

Face of the Country. Somerset, and parts of Fayette, Westmoreland, Cambria, Indiana, Jefferson and McKean are mountainous, with intervening vallies of rich, arable land. The hilly portions of Washington, and portions of Fayette, Westmoreland, and Alleghany counties are fertile, with narrow vales of rich land intervening. The hills are of various shapes, and heights, and the ridges are not uniform, but pursue various and different directions. North of Pittsburgh, the country is hilly and broken, but not mountainous, and the bottom lands on the water courses are wider and more fertile. On French creek, and other branches of the Alleghany river there are extensive tracts of rich bottom, or intervale lands, covered with beech, birch, sugar maple, pine, hemlock, and other trees common to that portion of the United States. The pine forests in Pennsylvania and New York, about the heads of the Alleghany river, produce vast quantities of lumber, which are sent annually to all the towns along the Ohio and Mississippi rivers. It is computed that not less than thirty million feet of lumber are annually sent down the Ohio from this source.

Soil, Agriculture, &c. Portions of the country are excellent for farming. The *glade* lands, as they are called, in Greene and other counties, produce oats, grass, &c., but are not so good for wheat and corn. Those counties which lie towards lake Erie are better adapted to grazing. Great numbers of cattle are raised here. Washington and other counties south of Pittsburgh produce great quantities of wool. The Monongahela has been famous for its whisky, but it is gratifying to learn that it is greatly on the decline, and that its manufacture begins to be regarded as it should be,—ruinous to society. A large proportion of the distilleries are reported to have been abandoned. Bituminous coal abounds in all the hills around Pittsburgh, and over most parts of Western Pennsylvania. Iron ore is found abundantly in the counties along the Alleghany, and many furnaces and forges are employed in its manufactory. Salt springs abound on the Alleghany, and especially on the Conemaugh and Kiskiminitas, where salt, in large quantities, is manufactured.

The natural advantages of Western Pennsylvania are great. Almost every knoll, hill and mountain can be turned to some good account, and its rivers, canals, rail and turnpike roads afford facilities for intercommunication, and for transportation of the productions to a foreign market. The advantages of this region for trade, agriculture, raising stock, and manufacturing, are great. The streams furnish abundant mill-seats, the air is salubrious, and the morals of the community good. Till recently, Pennsylvania has been neglectful to provide for common schools. A school system is now in successful operation, and has a strong hold on the confidence and affections of the people in this part of the State.

Internal Improvements. Pennsylvania has undertaken an immense system of internal improvements, throughout the State. The Alleghany portage rail-road commences at Hollidaysburgh, on the Juniata river, at the termination of the eastern division of the great Pennsylvania canal, and crosses the Alleghany ridge at Blair's Gap, summit thirty-seven miles, to Johnstown on the Conemaugh. Here it connects with the western division of the same canal. It ascends and descends the mountain by five inclined planes on each side, overcoming in ascent and descent 2570 feet, 1398 of which are on the eastern, and 1172 on the western side of the mountain. Five hundred and sixty-three feet are overcome by grading, and 2007 feet by the planes. On this line, also, are four extensive viaducts, and a tunnel eight hundred and seventy feet long, and twenty feet wide, through the staple bend of the Conemaugh river. The western division of the Pennsylvania canal commences at Johnstown, on the Conemaugh, pursues the course of that stream, and also that of the Kiskiminitas and Alleghany rivers, and finally terminates at Pittsburgh. In its course from Johnstown, it passes through the towns of Fairfield, Lockport, Blairsville, Saltzburg, Warren, Leechburg, and Freeport, most of which are small villages, but increasing in size and business. "The canal is one hundred and four miles in length: lockage four hundred and seventy-one feet, sixty-four locks (exclusive of four on a branch canal to the Alleghany), ten dams, one tunnel, sixteen aqueducts, sixty-four culverts, thirty-nine waste-wiers, and one hundred and fifty-two bridges.

"The canal commissioners, in their reports to the legislature, strongly recommend the extension of this division to the town of Beaver, so as to unite with the Beaver division. By a recent survey, the distance was ascertained to be 25.065 miles, and the estimated cost of construction, $263,821. This, with a proposed canal from Newcastle to Akron, on the Ohio and Erie canal, will form a continuous inland communication between Philadelphia and New Orleans, of 2435 miles, with the exception of the passage over the Alleghany portage rail-road, of 36.69 miles in length.* It is three hundred and ninety-five miles from Philadelphia to Pittsburgh by this canal.

The Beaver division of the Pennsylvania canal commences at the town of Beaver, on the Ohio river, at the junction of the Big Beaver river, twenty-five and a half miles below Pittsburgh, ascends the valley of that river, thence up the Chenango creek to its termination in Mercer county, a distance of 42.68 miles. This work, together with a feeder on French creek, and other works now in progress, are parts of a canal intended eventually to connect the Ohio river with lake Erie, at the town of Erie; which, when finished, will probably be about one hundred and thirty miles in length. It is also proposed to construct a canal from Newcastle, on the Beaver division, 24.75 miles above the town of Beaver, along the valley of the Mahoning river, to Akron, near the portage summit of the

* See "Mitchell's Compendium of the Internal Improvements in the United States," where much valuable information of the rail-roads and canals of the United States is found in a small space.

Ohio and Erie canal, eighty-five miles in length, eight miles of which are in Pennsylvania, and the residue in Ohio. Estimated cost, $764,372.

The Cumberland, or national road, crosses the south-western part of Pennsylvania. It passes through Brownsville where it crosses the Monongahela river, and Washington, into a corner of Virginia to Wheeling, where it crosses the Ohio river, and from thence through Ohio, Indiana and Illinois to the Mississippi river, or perhaps to the western boundary of Missouri.

Chief Towns. Brownsville, situated on the east side of the Monongahela river, is in a romantic country, surrounded with rich farms and fine orchards, and contains about 1200 inhabitants. It is at the head of steamboat navigation. *Washington* is the county seat of Washington county, surrounded with a fertile but hilly country, contains about 2000 inhabitants, and has a respectable college. *Cannonsburgh* is situated on the west side of Chartier's creek, eight miles north of Washington. It also has a flourishing college, with buildings, in an elevated and pleasant situation. *Uniontown* is the county seat of Fayette, on the National road, and contains about 1500 inhabitants. *Greensburg* is the seat of justice for Westmoreland county, on the great turnpike road from Philadelphia by Harrisburgh to Pittsburgh, and has about 850 inhabitants. *Beaver* is situated at the mouth of Big Beaver, on the Ohio, with a population of 1000 or 1200, and is a place of considerable business. *Meadville* is the seat of justice for Crawford county, situated near French creek, and has about 1200 inhabitants. Here is a college established by the Rev. Mr. Alden, some years since, to which the late Dr. Bentley, of Salem, Mass., bequeathed a valuable library. It is now under the patronage of the Methodist Episcopal church.

Erie is a thriving town, situated on the south side of lake Erie, one hundred and twenty miles north of Pittsburgh. Steam-boats that pass up the lake from Buffalo, usually stop here, from whence stage routes communicate with Pittsburgh, and many other towns in the interior. The portage from this place to the navigable waters of the Alleghany river is fifteen miles, over a turnpike road. The population of Erie is from 1500 to 2000, and increasing.

Waterford, the place where the Erie portage terminates, is situated on the north bank of the French creek; it is a place of considerable business. French creek is a navigable branch of the Alleghany river. *Franklin*, *Kittanning* and *Freeport* are respectable towns on the Alleghany river, between Pittsburgh and Meadville.

Economy is the seat of the German colony, under the late Mr. Rapp, which emigrated from their former residence of Harmony, on the Wabash river, in Indiana. It is a flourishing town on the right bank of the Ohio, eighteen miles below Pittsburgh. It has several factories, a large church, a spacious hotel, and eight or nine hundred inhabitants, living in a community form, under some singular regulations. The Economists or Harmonists, as they were called, in Indiana, are an industrious, moral and enterprising community, with some peculiarities in their religious notions. There are many other towns and villages in Western Pennsylvania, of moral, industrious inhabitants, which the limits of this work will not permit me to notice.

PITTSBURGH is the emporium of Western Pennsylvania, and from its manufacturing enterprise, especially in iron wares, has been denominated the "Birmingham of the West." It stands on the land formed at the junction of the Monongahela and Alleghany rivers, on a level alluvion deposit, but entirely above the highest waters, surrounded with hills. This place was selected as the site of a fort and trading depôt by the French, about eighty years since, and a small stockade erected, and called Fort du Quesne, to defend the occupancy of it by the English, and to monopolize the Indian trade. It came into the possession of the British, upon the conquest of this country, after the disastrous defeat of Gen. Braddock; and under the administration of the elder Pitt, a fort was built here under the superintendence of Lord Stanwix, that cost more than $260,000, and called Fort

Pitt. In 1760, a considerable town arose around the fort, surrounded with beautiful gardens and orchards, but it decayed on the breaking out of the Indian war, in 1763. The origin of the present town may be dated 1765. Its plan was enlarged, and re-surveyed in 1784, and then belonged to the Penn family, as a part of their hereditary manor. By them it was sold.

The Indian wars in the West, retarded its growth for several years after, but since, it has steadily increased, according to the following table:—

1800,	1,565
1810,	4,768
1820,	7,248
1830,	12,542
1835, estimated,	30,000

The estimate of 1835 includes the suburbs. The town is compactly built, and some streets are handsome; but the use of coal for culinary and manufacturing purposes, gives the town a most dingy and gloomy aspect. Its salubrity and admirable situation for commerce and manufactures ensure its future prosperity and increase of population. The exhaustless beds of coal in the bluffs of the Monongahela, and of iron ore, which is found in great abundance in all the mountainous regions of Western Pennsylvania, give it preëminence over other western cities, for manufacturing purposes. It really stands at the head of steam-boat navigation, on the waters of the Ohio; for the Alleghany and Monongahela rivers are navigable only at high stages of water, and, by the recent improvements in the channel of the Ohio and the use of light draft boats, the navigation to Pittsburgh is uninterrupted, except in winter.

The suburbs of Pittsburgh, are Birmingham, on the south bank of the Monongahela, Alleghany town, on the opposite side of the Alleghany river, and containing a population of about 7000, Lawrenceville, Northern and Eastern Liberties.

Manufactures.

Nail Factories and Rolling Mills.	Weight.	Value.
Union, ...	720,000	$43,200
Sligo, ...	400,000	32,000
Pittsburgh, ...	782,887	86,544
Grant's hill, ...	500,000	20,000
Juniata, ...	500,000	30,000
Pine Creek, ...	457,000	34,100
Miscellaneous factories, ...	360,000	28,200

The foregoing table was constructed in 1831. Doubtless this branch of business has greatly increased.

The same year there were twelve foundries in and near Pittsburgh, which converted 2963 tons of metal into castings, employed 132 hands, consumed 87,000 bushels of charcoal, and produced the value of $189,614.

The following sketch of manufactures in Pittsburgh and vicinity, is copied from Tanner's Guide, published in 1832:—

Steam engines, thirty-seven, which employed 123 hands. Value, $180,400.

Cotton factories, eight, with 369 power-looms, 598 hands; value, $300,134. In the counties of Westmoreland and Alleghany, there are five cotton factories.

In Pittsburgh and the two counties just named, are eight paper mills, valued at $165,000.

In Pittsburgh and vicinity are five steam mills, which employ 50 hands. Value of their products annually, $80,000.

There are five brass-foundries and eight coppersmiths' shops. Value of the manufactures, $25,000.

Within the limits of the city, there are thirty blacksmiths' shops, which employ 136 hands. There are also four gunsmiths, and nine silversmiths and watch repairers.

In Pittsburgh, and the counties of Westmoreland and Alleghany, there are twenty-six saddleries and forty-one tanneries, sixty-four brick-yards and eleven potteries. There are in the city, four breweries, and four white lead manufactories, at which 7400 kegs are made annually; value, $27,900.

There are six printing offices in Pittsburgh, and six more in the two counties.

The estimated value of the manufactures of every kind in Pittsburgh, and the counties of Alleghany and Westmoreland, in 1831, was $3,978,469. Doubtless they have greatly increased since.

Coal. The bituminous coal formations around Pittsburgh, are well deserving the attention of geologists. Coal hill, on the west side of the Monongahela, and immediately opposite Pittsburgh, is the great source of this species of fuel; and the miners, in some places, have perforated the hill to the distance of several hundred feet. It is found in strata from six inches to ten or twelve feet in thickness, and often at the height of three hundred feet above the bed of the river, in the hills around Pittsburgh, and along the course of the Alleghany and Monongahela. Below this one stratum, which is of equal elevation, none is found till you reach the base of the hill, below the bed of the river. Besides supplying Pittsburgh, large quantities are sent down the river.

There are in Pittsburgh (or were two years since), three Baptist churches, or congregations, one of which is of Welch; four Presbyterian, four Methodist, one Episcopal, one Roman Catholic (besides a cathedral on Grant's hill), one Covenanter, one Seceder, one German Reformed, one Unitarian, one Associate Reformed, one Lutheran, one African, and perhaps some others in the city or suburbs.

Of the public buildings deserving notice, I will name the *Western University of Pennsylvania*, which stands on the Monongahela, near Grant's hill; the *Penitentiary*, in Alleghany town, which has cost the State an immense amount, and is conducted on the principle of solitary confinement; the *Presbyterian Theological Seminary* is also in Alleghany town;

the *Museum;* the *United States Arsenal*, about two miles above the city, at Lawrenceville. It encloses four acres, and has a large depôt for ordnance, arms, &c. The *City Water-works* is a splendid monument of municipal enterprise. The water is taken from the Alleghany river, by a pipe, fifteen inches in diameter, and carried 2439 feet, and one hundred and sixteen feet elevation, to a reservoir on Grant's hill, capable of receiving 1,000,000 gallons. The water is raised by a steam engine of eighty-four horse power, and will raise 1,500,000 gallons in twenty-four hours. The aqueduct of the Pennsylvania canal, across the Alleghany river, is also deserving attention.

The inhabitants of Pittsburgh are a mixture of English, French, Scotch, Irish, German and Swiss artizans and mechanics, as well as of native born Americans, who live together in much harmony. Industry, sobriety, morality and good order generally prevail. Extensive revivals of religion prevailed here a year or two since.

The population of Western Pennsylvania is characterized for industry, frugality, economy and enterprise. Temperance principles have made considerable progress of late years.

CHAPTER VIII.

WESTERN VIRGINIA.

Sulphur, Hot and Sweet Springs—Chief Towns.

WESTERN VIRGINIA embraces all that part of Virginia that lies upon the western waters. The counties are Brooke, Ohio, Monongalia, Harrison, Randolph, Russell, Preston, Tyler, Wood, Greenbrier, Kenawha,* Mason, Lewis, Nicholas, Logan, Cabell, Monroe, Pocahontas, Giles, Montgomery, Wythe, Grayson, Tazewell, Washington, Scott, and Lee,—26.

Its principal river is the Kenawha and its tributaries. Of these, Gaula, New river and Greenbrier are the principal. New river is the largest, and rises in North Carolina. The Monongahela drains a large district; the little Kenawha, Guyandotte and Sandy are smaller streams. The latter separates Virginia from Kentucky, for some distance.

Much of Western Virginia is mountainous,

* I have adopted the orthography of the legislature.

lying in parallel ridges, which are often broken by streams. Some of the vallies are very fertile. The Kenawha valley is narrow, but extends to a great distance. The salt-manufactories extend from Charlestown, up the Kenawha, the distance of twelve miles. They are twenty in number, and manufacture nearly two millions of bushels annually. The river is navigable for steam-boats to this point, at an ordinary depth of water. Coal is used in the manufactories, which is dug from the adjacent mountains, and brought to the works on wooden rail-ways. Seven miles above Charlestown is the famous burning spring. Inflammable gas escapes, which, if ignited, will burn with great brilliancy for many hours, and even for several days, in a favorable state of the atmosphere. The State of Virginia has constructed a tolerably good turnpike road from the mouth of the Guyandotte, on the Ohio, to Staunton. It passes through Charlestown, and along the Kenawha river to the falls; from thence it extends along the course of New river, and across Sewall's mountain, by Louisburg, to Staunton. The falls of Kenawha are in a romantic region, and merit the attention of the traveler. Marshall's pillar is a singular projecting rock, that overhangs New river, 1015 feet above its bed. The stage road passes near its summit.

This route is one of the great stage routes leading from the Ohio valley to Washington city, and to all parts of old Virginia.

The *White Sulphur*, *Red Sulphur*, *Hot*, *Warm*, and *Sweet Springs* are in the mountainous parts of Virginia, and on this route. These are all celebrated as watering-places, but the White Sulphur spring is the great resort of the fashionable of the Southern States. Let the reader imagine an extensive camp-ground, a mile in circumference, the camps neat cottages, built of brick, or framed, and neatly painted. In the centre of this area are the springs, bath-houses, dining-hall, and mansion of the proprietor. The cottages are intended for the accommodation of families, and contain two rooms each. This is by far the most extensive watering-place in the Union. Of the effect of such establishments on *morals* I shall say nothing. The reader will draw his own conclusions, when he understands that the card-table, roulette, wheel of fortune and dice-box are amongst its principal amusements. Here, not unfrequently, cotton bales, negroes, and even plantations, change owners in a night. The scenery around is highly picturesque and romantic. Declivities and mountains, sprinkled over with evergreens, are scattered in wild confusion. A few miles from White Sulphur springs, you pass the dividing line,—the Alleghany ridge,—and pass from Western into Middle Virginia.

Chief Towns. Wheeling is the principal commercial town, and a great thoroughfare, in Western Virginia. It has a large number of stores and commission ware-houses; and contains six or eight thousand inhabitants. It is ninety-two miles by water, and fifty-five miles by land, from Pittsburgh. It has manufactures of cotton, glass and earthen ware: boats are built here. The Cumberland or national road crosses the Ohio at this place, over which a bridge is about to be erected. The town is surrounded with bold, precipitous hills, which contain inexhaustible quantities of coal. At extreme low water steam-boats ascend no higher than Wheeling.

Charlestown, Wellsburgh, Parkersburgh, Point Pleasant, Clarksburgh, Abington, Louisburg, and many others, are pleasant and thriving towns.

The climate of Western Virginia is preëminently salubrious. The people, in their manners, have considerable resemblance to those of Western Pennsylvania. There are fewer slaves, less wealth, more industry and equality, than in the "old dominion," as Eastern Virginia is sometimes called.

CHAPTER IX.

MICHIGAN.

Extent—Situation—Boundaries—Face of the Country—Rivers, Lakes, &c.--Soil and Productions--Subdivisions, Counties—-Chief Towns—Education—-Improvements projected—Boundary Dispute—Outline of the Constitution.

MICHIGAN is a large triangular peninsula, surrounded on the east, north and west, by lakes, and on the south by the States of Ohio and Indiana. Lake Erie, Detroit river, lake St. Clair and St. Clair river, lie on the east for 140 miles; lake Huron on the north-east and north, the straits of Mackinaw on the extreme north-west, and lake Michigan on its western side. Its area is about 40,000 square miles.

Face of the Country. Its general surface is level, having no mountains, and no very elevated hills. Still, much of its surface is undulating, like the swelling of the ocean. Along the shore of lake Huron, in some places, are high, precipitous bluffs, and along the eastern shore of Michigan are hills of pure sand, blown up by the winds from the lake. Much of the country bordering on lakes Erie, Huron, and St. Clair, is level,—somewhat deficient in good water, and for the most part heavily timbered. The interior is more undulating, in some places rather hilly, with much fine timber, interspersed with oak "openings," "plains," and "prairies."

The *plains* are usually timbered, destitute of undergrowth, and are beautiful. The soil is rather gravelly. The *openings* contain scattering timber in groves and patches, and resemble those tracts called *barrens* farther south. There is generally timber enough for farming purposes, if used with economy, while it costs but little labor to clear the land. For the first ploughing, a strong team of four or five yoke of oxen is required, as is the case with prairie.

The *openings* produce good wheat.

The *prairies*, will be described more particularly under the head of Illinois. In Michigan, they are divided into wet and dry. The latter possess a rich soil, from one to four feet deep, and produce abundantly all kinds of crops common to 42 degrees of north latitude, especially those on St. Joseph river. The former afford early pasturage for emigrants, hay to winter their stock, and, with a little labor, would be converted into excellent artificial meadows. Much of the land that now appears wet and marshy, will, in time, be drained, and be the first rate soil for farming.

A few miles back of Detroit is a flat, wet country, for considerable extent, much of it heavily timbered,—the streams muddy and sluggish,—some wet prairies,—with dry, sandy ridges intervening. The timber consists of all the varieties found in the Western States; such as oaks of various species, walnut, hickory, maple, poplar, ash, beech, &c., with an intermixture of white and yellow pine.

Rivers and Lakes. In general, the country abounds with rivers and small streams. They rise in the interior, and flow in every direction to the lakes which surround it. The northern tributaries of the Maumee rise in Michigan, though the main stream is in Ohio, and it enters the west end of lake Erie on the "debatable land." Proceeding up the lake, Raisin and then Huron occur. Both are navigable streams, and their head waters interlock with Grand river, or Washtenong, which flows into lake Michigan. River Rouge enters Detroit river, a few miles below the city of Detroit. Raisin rises in the county of Lenawee, and passes through Monroe. Huron originates amongst the lakes of Livingston, passes through Washtenaw, and a corner of Wayne, and enters lake Erie towards its north-western corner. Above Detroit is river Clinton, which heads in Oakland county, passes through Macomb, and enters lake St.

Clair. Passing by several smaller streams, as Belle, Pine, and Black rivers which fall into St. Clair river, and going over an immense tract of swampy, wet country, between lake Huron and Saginaw bay, in Sanilac county, we come to the Saginaw river. This stream, is formed by the junction of the Tittibawassee, Hare, Shiawassee, Flint, and Cass rivers, all of which unite in the centre of Saginaw county, and form the Saginaw river, which runs north, and enters the bay of the same name. The Tittibawassee rises in the country west of Saginaw bay, runs first a south, and then a south-eastern course, through Midland county into Saginaw county, to its junction. Pine river is a branch of this stream, that heads in the western part of Gratiot county, and runs north-east into Midland. Hare, the original name of which is Waposebee, commences in Gratiot and the north-west corner of Shiawassee counties, and runs an east and north-east course. The heads of the Shiawassee, which is the main fork of the Saginaw, are found in the counties of Livingston and Oakland. Its course is northward. Flint river rises in the south part of Lapeer county, and runs a north-western course some distance past the centre of the county, when it suddenly wheels to the south, then to the west, and enters Genesee county, through which it pursues a devious course towards its destination. Cass river rises in Sanilac county, and runs a western course. These rivers are formed of innumerable branches, and water an extensive district of country. Other smaller streams enter lake Huron, above Saginaw bay; but the whole country across, to lake Michigan is yet a wilderness, and recently purchased of the Indians. Doubtless it will soon be surveyed and settled. On the western side of the State are Traverse, Ottawa, Betsey, Manistic, Pent, White, Maskegon, Grand, Kekalamazoo, and St. Joseph, all of which fall into lake Michigan. Those above Grand river are beyond the settled portion of the State. Grand river is the largest in Michigan, being 270 miles in length, its windings included. Its head waters interlock with the Pine, Hare, Shiawassee, Huron, Raisin, St. Joseph and Kekalamazoo. A canal project is already in agitation to connect it with the Huron, and open a water communication from lake Erie, across the peninsula, direct to lake Michigan. Grand river is now navigable for batteaux, 240 miles, and receives in its course, Portage, Red Cedar, Lookingglass, Maple, Muscota, Flat, Thorn-Apple, and Rouge rivers, besides smaller streams. It enters lake Michigan 245 miles south-westerly from Mackinaw, and 75 north of St. Joseph;—is between fifty and sixty rods wide at its mouth, with eight feet water over its bar. Much of the land on Grand river and its tributaries, is excellent, consisting of six or seven thousand square miles;—and, considering its central position in the State,—the general fertility of its soil,—the good harbor at its mouth,—the numerous mill sites on its tributaties,—this region may be regarded as one of the most interesting portions of Michigan. The Kekalamazoo rises in Jackson and Eaton counties, passes through Calhoun and the northern part of Kalamazoo, enters the south-eastern part of Allegan, and passes diagonally through it to the lake. There is much first-rate land, timber, prairie and openings on its waters, and is rapidly settling.

The St. Joseph country is represented by some as the best country in Michigan. This stream has several heads in Branch, Hillsdale, Jackson, Calhoun, and Kalamazoo counties, which unite in St. Joseph county, through which it passes diagonally to the south-west, into Indiana,—thence through a corner of Elkhart county, into St. Joseph of that State, makes the "South Bend," and then runs north-westerly into Michigan, through Berrian county, to the lake. The town of St. Joseph is at its mouth. It has Pigeon, Prairie, Hog, Portage Christianna, Dowagiake, and Crooked rivers for tributaries, all of which afford good mill sites. In Cass and St. Joseph counties, are Four-mile, Beardsley, Townsend, McKenny, LaGrange, Pokagon, Young, Sturges, Nottawa-Sepee, and White Pigeon prairies, which are rich tracts of country, and fast filling up with inhabitants.

Michigan abounds with small lakes and ponds. Some have marshy and unhealthy borders; others are transparent fountains, surrounded with beautiful groves, an undulating country, pebbly and sandy shores, and teeming with excellent fish. The counties of Oakland, Livingston, Washtenaw, Jackson, Barry and Kalamazoo are indented with them.

Productions. These are the same, in general, as those of Ohio and New York. Corn and wheat grow luxuriantly here; rye, oats, barley, buckwheat, potatoes and all the garden vegetables common to the climate, grow well; all the species of grasses are produced luxuriantly; apples, and other fruit, abound in the older settlements, especially among the French about Detroit. It will be a great fruit country.

Subdivisions. Michigan had been divided into thirty-three counties in 1835, some of which were attached to adjacent counties for judicial purposes. Other counties may have been formed since. The following organized counties show the population of the State (then Territory), at the close of 1834:

Counties.	Population.	Seats of Justice.	Dist. from Detroit.
Berrian,	1,787	Berrian,	180
Branch,	764	Branch,	133
Calhoun,	1,714	Eckford,	100
Cass,	3,280	Cassopolis,	160
Jackson,	1,865	Jacksonburgh,	77
Kalamazoo,	3,124	Bronson,	137
Lenawee,	7,911	Tecumseh,	63
Macomb,	6,055	Mount Clemens,	25
Monroe,	8,542	Monroe,	36
Oakland,	13,844	Pontiac,	26
St. Clair,	2,244	St. Clair,	60
St. Joseph,	3,168	White Pigeon,	135
Washtenaw,	14,920	Ann Arbor,	42
Wayne,	16,638	Detroit,	
Total,	85,856		

The other counties are Hillsdale, Van Buren, Allegan, Barry, Eaton, Ingham, Livingston, Lapeer, Genesee, Shiawassee, Clinton, Ionia, Kent, Ottawa, Oceana, Gratiot, Isabella, Midland, Saginaw, Sanilac, Gladwin and Arenac, the population of which are included in the counties given in the table. Doubtless, the population of Michigan exceeds one hundred thousand.

The counties are subdivided into incorporated townships, for local purposes, the lines of which usually correspond with the land surveys.

For the sales of public lands, the State is divided into three land districts, and land-offices are established at Detroit, Monroe, and Bronson.

Chief Towns. Detroit is the commercial and political metropolis. It is beautifully situated on the west side of the river Detroit, eighteen miles above Malden, in Canada, and eight miles below the outlet of lake St. Clair. A narrow street, on which the wharves are built, runs parallel with the river. After ascending the bench or bluff, is a street called Jefferson avenue, on which the principal buildings are erected. The older dwellings are of wood, but many have been recently built of brick, with basements of stone, the latter material being brought from Cleveland, Ohio. The primitive forest approaches near the town: the table land extends twelve or fifteen miles interior, when it becomes wet and marshy. Along Detroit river the ancient French settlements extend several miles, and the inhabitants exhibit all the peculiar traits of the French on the Mississippi. Their gardens and orchards are valuable.

The public buildings of Detroit are a state-house, a council-house, an academy, and two or three banking-houses. There are five churches for as many different denominations, in which the Episcopalians, Presbyterians, Baptists, Methodists and Roman Catholics worship. The Catholic congregation is the largest, and they have a large cathedral. Stores and commercial warehouses are numerous, and business is rapidly increasing. Town lots, rents, and landed property in the vicinity, are rising rapidly. Lots have advanced, within two or three years, in the business parts of the city, more than one thousand per cent. Mechanics of all descriptions, and particularly those in the building line, are much wanted here, and in other towns in Michigan. The population is supposed to be about 10,000, and is rapidly increasing. This place commands the trade of all the upper lake country.

Monroe, the seat of justice for Monroe county, is situated on the right bank of the river Raisin, opposite the site of old Frenchtown. Three years since, it had about one hundred and fifty houses, of which twenty or thirty were of stone, and 1600 inhabitants. There were also two flouring and several saw-mills, a woollen factory, an iron foundry, a chair factory, &c., and an abundant supply of water power. The "Bank of the River Raisin," with a capital of $100,000, is established here. The Presbyterians, Episcopalians, Baptists, Methodists and Roman Catholics have houses of worship and ministers here.

It was at this place, or rather at Frenchtown in its vicinity, that a horrible massacre of American prisoners took place during the last war with Great Britain, by the Indians under Gen. Proctor. The sick and wounded were burned alive in the hospital, or shot as they ran shrieking through the flames! Of the seven hundred young men barbarously murdered here, many were students at law, young physicians and merchants, the best blood of Kentucky.

Mount Clemens, Brownstown, Ann Arbor, Pontiac, White Pigeon, Tecumseh, Jacksonburgh, Niles, St. Joseph, Spring Arbor, and many others, are pleasant villages, and will soon become populous.

Education. Congress has made the same donation of lands, as to other Western States, and will, doubtless, appropriate the same per centage on the sales of all public lands, when the State is admitted into the Union, as has been appropriated in the other new States. A respectable female academy is in operation at Detroit. The Presbyterian denomination are about establishing a college at Ann Arbor, the

Methodists a seminary at Spring Arbor, the Baptists one in Kalamazoo county, and the Roman Catholics, it is said, have fixed their post at Bertrand, a town on the St. Joseph river, in the south-eastern corner of Berrian county, and near to the boundary line of Indiana. Much sentiment and feeling exists in favor of education and literary institutions, among the people.

Projected Improvements. A survey has been made for a rail-road across the peninsula of Detroit, through the counties of Wayne, Washtenaw, Jackson, Calhoun, Kalamazoo, Van Buren and Berrian, to the mouth of St. Joseph river. Another project is, to commence at or near Toledo, on the Maumee river, and pass through the southern counties of Michigan, into Indiana, and terminate at Michigan city. A third project is, to open a water communication from the navigable waters of Grand river to Huron river, and, by locks and slack-water navigation, enter lake Erie. A canal from the mouth of Maumee bay to lake Michigan, has also been spoken of as a feasible project; or one from the mouth of the river Raisin to the St. Joseph would open a similar communication. It has also been suggested to improve the river Raisin by locks and slack-water navigation. Doubtless, not many years will elapse before some of these projects will prove realities.

Boundary Dispute. This unpleasant dispute between Ohio and Michigan, relates to a strip of country about fifteen miles in width at its eastern and seven miles at its western end, lying between the north-eastern part of Indiana and the Maumee bay. A portion of the Wabash and Erie canal, now constructing by Indiana, and which is dependent for its completion on either Ohio or Michigan, passes over this territory. Michigan claims it, by virtue of an ordinance of Congress, passed July 13, 1787, organizing the *North-Western Territory*, in which the boundaries of *three* States were laid off; "provided, that the boundaries of these three States shall be subject so far to be altered, that, if Congress shall hereafter find it expedient, they shall have authority to form one or two States in that part of the said Territory *which lies north of an east and west line drawn through the southerly bend or extreme of lake Michigan.*" Ohio claims it by possession; and because, by being received into the Union with this portion in possession, Congress virtually annulled that part of the former ordinance that fixed the south bend of lake Michigan as the boundary line, and by having run the line north of this.

Outlines of the Constitution. A convention assembled at Detroit, May 11, 1835, and framed a constitution for a State government, which was submitted to and ratified by vote of the people, on the first Monday in October.

The powers of the government are divided into three distinct departments; the legislative, the executive, and the judicial.

The legislative power is vested in a *Senate* and *House of Representatives.* The representatives are to be chosen annually; and their number cannot be less than forty-eight, nor more than one hundred.

The senators are to be chosen every two years, one half of them every year, and to consist, as nearly as may be, of one third of the number of the representatives.

The census is to be taken in 1837 and 1845, and every ten years after the latter period, and also after each census taken by the United States: the number of senators and representatives is to be apportioned anew among the several counties, according to the number of white inhabitants.

The legislature is to meet annually, on the first Monday in January.

The executive power is to be vested in a governor, who holds his office for two years. Upon a vacancy, the lieutenant governor performs executive duties. The first election was held on the first Monday in October, 1835; and the governor and lieutenant governor hold their offices till the first Monday in January, 1838.

The judicial power is vested in one *Supreme Court*, and in such other courts as the legislature may, from time to time, establish. The judges of the supreme court are to be appointed by the governor, with the advice and consent of the senate, for the term of seven years. Judges of all county courts, associate judges of circuit courts and judges of probate are to be elected by the people, for the term of four years.

Each township is authorized to elect four justices of the peace, who are to hold their offices for four years. In all elections, every white male citizen, above the age of twenty-one years, having resided six months next preceding any election, is entitled to vote at such election. Slavery, lotteries and the sale of lottery-tickets are prohibited.

The seat of government is to be at Detroit, or such other place or places as may be prescribed by law, until the year 1847, when it is to be permanently fixed by the legislature.

CHAPTER X.

OHIO.

Boundaries—Divisions—Face of the Country—Soil and Productions—Animals—Minerals—Financial Statistics—Canal Fund—Expenditures—Land Taxes—School Fund—Statistics—Canal Revenues—Population at different Periods—Rivers—Internal Improvements—Manufactures—Cities and Towns, Cincinnati, Columbus—Education—Form of Government—Antiquities—History.

Ohio is bounded on the north by lake Erie and the State of Michigan, east by Pennsylvania and the Ohio river, south by the Ohio river, which separates it from Virginia and Kentucky, and west by Indiana. The meanderings of the Ohio river extend along the line of this State four hundred and thirty-six miles. It is about two hundred and twenty-two miles in extent, both from north to south and from east to west. After excluding a section of lake Erie, which projects into its northern borders, Ohio contains about 40,000 square miles, or 25,000,000 acres of land.

Divisions. Nature has divided this State

into four departments, according to its principal waters.

1. The lake country, situated on lake Erie, and embracing all its northern part. Its streams all run into the lake, and reach the Atlantic ocean through the gulf of St.'Lawrence.

2. The Muskingum county, on the eastern side, and along the river of that name.

3. The Scioto country, in the middle; and,

4. The Miama country, along the western side.

For civil purposes, the State is divided into seventy-five counties, and these are again subdivided into townships. Their names, date of organization, number of square miles, number of organized townships, seats of justice, and bearing and distance from Columbus, are exhibited in the following table:

Counties.	When organized.	Square miles.	Number of Townships.	Seats of Justice.	Bearing and distance frm Columbus.
Adams, . . .	1797	550	10	West Union, . . .	101 *s.*
Allen, ...	1831	542		Lima, ...	110 *n.w.*
Ashtabula,. .	1811	700	27	Jefferson, ...	200 *n.w.*
Athens, ...	1805	740	19	Athens, ...	73 *s.e.*
Belmont, . .	1801	536	16	St. Clairsville,. . .	116 *e.*
Brown, ...	1818	470	14	Georgetown,. . . .	104 *s.*
Butler, ...	1803	480	13	Hamilton, ...	101 *s.w.*
Carrol,* . . .	1833	*	*	Carrolton, ...	125 *e.n.e.*

* Carroll county has been formed from Columbiana, Harrison, Stark and Tuscarawas, since the edition of the Ohio Gazetteer of 1833 was published, from which the foregoing table has been constructed. Hence the townships are not given.

Counties.	When organized.	Square miles.	Number of Townships.	Seats of Justice.	Bearing and distance frm Columbus.
Champaign, .	1805	417	12	Urbanna, ...	50 *w.n.w.*
Clark, ...	1818	412	10	Springfield, ...	44 *w.*
Clermont, . .	1800	515	12	Batavia, ...	96 *s.w.*
Clinton, ...	1810	400	8	Wilmington, ...	60 *s.w.*
Columbiana,	1803	*	*	New Lisbon, ...	150 *e.n.e.*
Coshocton, .	1811	562	21	Coshocton, ...	68 *n.e.*
Crawford, . .	1826	594	12	Bucyrus, ...	60 *n.*
Cuyahoga, . .	1810	475	19	Cleveland, ...	140 *n.n.e.*
Dark, ...	1817	660	10	Greenville, ...	93 *w.*
Delaware, .	1808	610	23	Delaware, ...	24 *n.*
Fairfield, . .	1800	540	14	Lancaster, ...	28 *s.e.*
Fayette, . . .	1810	415	7	Washington, ...	38 *s.w.*
Franklin, . .	1803	520	18	COLUMBUS, . . .	
Gallia, ...	1803	500	15	Gallipolis, ...	102 *s.s.e.*
Geauga, . . .	1805	600	23	Chardon, ...	157 *n.e.*
Greene, ...	1803	400	8	Xenia, ...	56 *w.s.w.*
Guernsey, .	1810	621	19	Cambridge, ...	76 *e.*
Hamilton, . .	1790	400	14	Cincinnati, ...	110 *s.w.*
Hancock, . .	1828	576	5	Findlay, ...	90 *n.n.w.*
Hardin, . . .	1833	570	—	Kenton, ...	70 *n.n.w.*
Harrison, . . .	1813	*—	13	Cadiz, ...	124 *e.n.e.*
Henry, ...	——	744	2	Napoleon, ...	161 *n.w.*
Highland, . .	1805	555	11	Hillsborough, . . .	62 *s.s.w.*
Hocking, . .	1818	432	9	Logan, ...	46 *s.s.e.*
Holmes, . . .	1825	422	14	Millersburg, ...	81 *n.e.*
Huron, ...	1815	800	29	Norwalk, ...	106 *n.*
Jackson, . . .	1816	490	13	Jackson, ...	73 *s.s.e.*
Jefferson, . .	1797	400	13	Steubenville, . . .	147 *e.n.e.*
Knox, ...	1808	618	24	Mount Vernon, . .	47 *n.n.e.*
Lawrence, . .	1817	430	12	Burlington, ...	130 *s.s.e.*
Licking, . . .	1808	666	25	Newark, ...	33 *e.n.e.*
Logan, ...	1818	425	9	Bellefontaine, . . .	50 *n.w.*
Lorain, ...	1824	580	19	Elyria, ...	130 *n.n.e.*
Lucas,*	1835		—	Toledo, ...	150 *n.n.w.*

* Lucas county has been recently formed from parts taken from Sandusky and Wood counties, and from the disputed country claimed by Michigan.

Counties.	When organized.	Square miles.	Number of Townships.	Seats of Justice.	Bearing and distance frm Columbus.
Madison, ..	1810	480	10	London, ...	25 *w.s.w.*
Marion, ...	1824	527	15	Marion, ...	45 *n.*
Medina, ...	1818	475	14	Medina, ...	110 *n.n.e.*
Meigs, ...	1819	400	12	Chester, ...	94 *s.s.e.*
Mercer, ...	1824	576	4	St. Mary's, ...	111 *n.w.*
Miami, ...	1807	410	12	Troy, ...	68 *n.of w.*
Monroe, ...	1815	563	18	Woodsfield, ...	120 *e.s.e.*
Montgomery,	1803	480	12	Dayton, ...	68 *w.*
Morgan, ...	1819	500	15	M'Connelsville, .	75 *s.e.*
Muskingum,	1804	665	23	Zanesville, ...	52 *e.*
Paulding,* ..	——	432	3		170 *n.w.*
Perry, ...	1818	402	12	Somerset, ...	46 *e.s.e.*
Pickaway, ..	1810	470	14	Circleville, ...	26 *s.*
Pike, ...	1815	421	9	Piketon, ...	64 *s.*
Portage, ...	1807	750	30	Ravenna, ...	135 *n.e.*
Preble, ...	1808	432	12	Eaton, ...	50 *w.*
Putnam,* ..		576	2		148 *n.w.*
Richland, ..	1813	900	25	Mansfield, ...	74 *n.n.e.*
Ross, ...	1798	650	16	Chilicothe, ...	45 *s.*
Sandusky, .	1820	600	10	Lower Sandusky,	105 *n.*
Scioto, ...	1803	700	14	Portsmouth, ...	90 *s.*
Seneca, ...	1824	540	11	Tiffin, ...	87 *n.*
Shelby, ...	1819	418	10	Sidney, ...	70 *n.w.*
Stark, ...	1809	*	16	Canton, ...	116 *n.e.*
Trumbull, .	1800	875	34	Warren, ...	160 *n.e.*
Tuscarawas,	1808	*	19	New Philadelphia,	100 *e.n.e.*
Union, ...	1820	450	9	Marysville.	30 *n.w.*
Vanwert,† ..		432			100 *n.w.*
Warren, ...	1803	400	9	Lebanon, ...	80 *s.w.*
Washington,	1788	713	19	Marietta, ...	106 *s.e.*
Wayne, ...	1812	660	20	Wooster, ...	89 *n.e.*
Williams, ..	1824	600	10	Defiance, ...	130 *n.w.*
Wood, ...	1820	750	7	Perrysburg, ...	135 *n.w.*

* Paulding, Putnam and Vanwert counties had not been organized at the period of our information.

† Much of the land in Vanwert is wet. The southern portion contains much swampy prairie.

There are nineteen congressional districts in Ohio, which elect as many members of Congress, and twelve circuits for courts of common pleas.

Face of the Country. The interior and northern parts of the State, bordering on lake Erie, are generally level, and, in some places, wet and marshy. The eastern and south-eastern parts, bordering on the Ohio river, are hilly and broken, but not mountainous. In some counties the hills are abrupt and broken; in others they form ridges, and are cultivated to their summits. Immediately on the banks of the Ohio, and other large rivers, are strips of rich alluvion soil.

The country along the Scioto and two Miami rivers, furnish more extensive bodies of rich, fertile land, than any other part of the State. The prairie land is found in small tracts near the head waters of the Muskingum and Scioto, and between the sources of the two Miami rivers, and especially in the north-western part of the State. Many of the prairies in Ohio are low and wet; some are elevated and dry, and exhibit the features of those tracts called "barrens," in Illinois. There are extensive plains, some of which are wet, towards Sandusky.

Soil and Productions. The soil, in at least three fourths of the State, is fertile; and some of it very rich. The poorest portion of Ohio is along the Ohio river, from fifteen to twenty-five miles in width, and extending from the national road, opposite Wheeling, to the mouth of the Scioto river. Many of the hills in this region are rocky.

Among the forest trees are oak of various species, white and black walnut, hickory, maple of different kinds, beech, poplar, ash of several kinds, birch, buckeye, cherry, chestnut, locust, elm, hackberry, sycamore, linden, with numerous others. Amongst the undergrowth are spice-bush, dogwood, iron-wood, pawpaw, hornbeam, black-haw, thorn, wild plum, grape vines, &c. The plains and wet prairies produce wild grass.

The agricultural productions are such as are common to the Eastern and Middle States. Indian corn, as in other Western States, is a staple grain, raised with much ease, and in great abundance: more than one hundred bushels are produced from an acre, on the rich alluvial soils of the bottom lands, though from forty to fifty bushels per acre ought to be considered an average crop. The State, generally, has a fine soil for wheat, and flour is produced for exportation in large quantities; rye, oats, buckwheat, barley, potatoes, melons, pumpkins and all manner of garden vegetables, are cultivated to great perfection. No markets in the United States are more profusely and cheaply supplied with meat and vegetables than those of Cincinnati and other large towns in Ohio. Hemp is produced to some extent, and the choicest kinds of tobacco are raised and cured in some of the counties east of the Muskingum river. Fruits of all kinds are raised in great plenty, especially apples, which grow to a large size, and are finely flavored. The vine and the mulberry have been introduced, and, with enterprise and industry, wine and silk might easily be added to its exports.

Animals. Bears, wolves and deer are still found in the forests and unsettled portions of the State. The domestic animals are similar to other States. Swine is one of the staple productions, and Cincinnati has been denominated the "pork market of the world." Other towns in the West, and in Ohio, are beginning to receive a share of this trade, especially along the lines of the Miami and the Erie canals. One hundred and fifty thousand hogs have been slaughtered and prepared for market in one season, in Cincinnati. About seventy-five thousand is the present estimated number, from newspaper authority. Immense droves of fat cattle are sent every autumn from the Scioto valley and other parts of the State. They are driven to all the markets of the east and south.

Minerals. The mineral deposits of Ohio, as yet discovered, consist principally of iron, salt, and bituminous coal, and are found, chiefly, along the south-eastern portion of the State. Let a line be drawn from the south-eastern part of Ashtabula county, in a south-western direction, by Northampton, in Portage county, Wooster, Mount Vernon, Granville, Circleville,, to Hillsborough, and thence south, to the Ohio river in Brown county, and it would leave most of the salt, iron and coal on the eastern and south-eastern side.

Financial Statistics. From the auditor's report to the legislature, January, 1836, the following items are extracted. The general revenue is obtained from moderate taxes on landed and personal property, and collected by the county treasurers, from insurance, bank and bridge companies, from lawyers, physicians, &c.

Collected in 1835, by the several county treasurers,	$150,080
Paid by banks, bridges, and insurance companies,	26,060
By lawyers and physicians,	1,598
From other sources,	24,028
Making an aggregate of	$201,766

The disbursements are:—

Amount of deficit, for 1834,	$16,622
Bills redeemed at the treasury, for the year ending Nov. 1835,	182,005
Interest paid on school fund,	33,101
Amounting to	$231,728
Showing a deficit in the revenue, of	$29,962

Canal Funds. These appear to be separate accounts from the general receipts and disbursements.

Miami Canal. The amount of money arising from the sales of esiami canal lands up to

Nov. 15, 1835, is $310,178. This sum has been expended in the extension of the canal north of Dayton.

Ohio Canal. The amount of taxes collected for canal purposes, for the year 1835, including tolls, sales of canal lands, school lands, balance remaining in the treasury of last year, &c., is $509,322. Only $38,242 of the general revenue were appropriated to canal purposes; of which, $35,507 went to pay interest on the school funds borrowed by the State.

The foreign debt is $4,400,000; the legal interest of which is $260,000 per annum. The domestic debt of the State, arising from investing the different school funds, is $579,287; the interest of which amounts to $34,757: making an aggregate annual interest paid by the State on loans, $294,757. The canal tolls for the year 1835 amounted to $242,357, and the receipts from the sale of Ohio canal lands, $64,549; making an aggregate income to the canal fund of $306,906 per annum;—a sum more than sufficient to pay the interest on all loans for canal purposes.

Items of Expenditure. Under this head the principal items of the expenditures of the State government are given.

Members, and officers of the General Assembly, per annum, ...	$43,987
Officers of government, ...	20,828
Keeper of the Penitentiary, ...	1,909
For new Penitentiary buildings, ...	46,050

State printing, ...	12,243
Paper and stationary, for use of the State,.	4,478
Certificates for wolf scalps, ...	2,824
Adjutant, and Quarter Master Generals, and Brigade Inspectors, ...	2,276
Treasurer's mileage, on settlement with the Auditor of State, ...	1,027
Deaf and Dumb Asylum, ...	5,700
Periodical works, &c., ...	400
Postage on documents, ...	545
Reporter to Court, in Bank, ...	300
Members and clerks of the Board of Equalization, and articles furnished, ...	1,960
Paymaster General,—Ohio militia, ...	2,000

The extra session of the legislature on the boundary line, in June, 1835, was $6,823.

Land Taxes. The amount of lands taxed, and the revenue arising therefrom, at several different periods, are herewith given, to show the progressive advance of the farming and other interests of the State.

Years.	Acres.	Taxes paid.
1809, ...	9,924,033	$63,991 87
1810, ...	10,479,029	67,501 60
1811, ...	12,134,777	170,546 74

From 1811 to 1816, the average increase of the taxes paid by the several counties, was $59,351. From 1816, the State rose rapidly in the scale of prosperity and the value of property. In 1820, the number of acres returned as taxable, exceeded a fraction of thirteen millions, while the aggregate of taxes, was $205,346.

The period of depression and embarrassment that followed throughout the West, prevented property from advancing in Ohio. In 1826, '27, '28, '29, '30, a material change in the amount of property taxable, took place, from a few hundred thousands to more than fifty millions. The total value of taxable property of the State, for 1835 (exclusive of three counties, from which returns had not been received), amounts to the sum of *ninety-four millions, four hundred and thirty-seven thousand, nine hundred and fifty-one dollars.*

School Funds. The amount of school funds loaned to the State, up to Nov. 15, 1835, is:—

Virginia military land fund,	$109,937
United States military land fund	90,126
Common school fund,	23,179
Athens university,	1,431
School section, No. 16,	453,000
Connecticut Western Reserve,	125,758
Total,	$803,432

The following tabular view of the acres of land, total amount of taxable property, and total amount of taxes paid for 1833, is taken from the Ohio Gazetteer. It should be noted, that in all the Western States, lands purchased of the government of the United States, are exempted from taxation for five years after sale. It is supposed that such lands are not included in the table. I have also placed the population of each county for 1830, from the census of that year; reminding the reader that great changes have since been made.

Counties.	Population 1830.	Acres of land.	Total amount of taxable property.	Total amount of taxes paid.
Adams, ...	12,231	234,822	$832,565	$6,995 41
Allen, ...	578	14,159	51,214	725 28
Ashtabula, .	14,584	449,742	1,347,900	13,524 97
Athens, ...	9,787	365,348	481,579	5,820 90
Belmont, ..	28,627	301,511	1,591,716	11,590 33
Brown, ...	17,867	267,130	1,458,944	8,179 35
Butler, ...	27,142	257,989	2,514,007	20,111 55
Carrol, ...		185,942	529,575	6,876 92
Champaign,.	12,131	233,493	908,571	5,956 66
Clark, ...	13,114	247,083	1,114,995	7,744 89
Clermont, ..	20,466	280,679	1,542,627	15,645 31
Clinton, ...	11,436	239,404	785,770	6,482 14
Columbiana,	35,592	317,796	1,491,099	14,217 28
Coshocton, .	11,161	246,123	850,708	9,307 28
Crawford, ..	4,791	79,582	217,675	3,630 09
Cuyahoga,..	10,373	292,252	1,401,591	18,122 96
Dark, ...	6,204	107,730	260,259	3,312 81
Delaware, .	11,504	338,856	831,093	8,516 66
Fairfield, ..	24,786	308,163	1,992,697	13,716 97
Fayette, ...	8,182	234,432	544,539	6,428 98
Franklin,...	14,741	325,155	1,663,315	13,247 34
Gallia, ...	9,733	205,727	427,962	4,826 55
Geauga, ...	15,813	381,380	1,427,869	15,832 65
Greene,....	14,801	251,512	1,441,907	12,082 36
Guernsey, .	18,036	275,652	908,109	9,855 72
Hamilton, ..	52,317	239,122	7,726,091	97,530 42
Hancock, ..	813	9,302	50,929	421 70
Harden, ...	210	125,607	118,425	1,291 43
Harrison, ..	20,916	22,412	1,025,210	12,400 97
Highland, ..	16,345	317,079	1,065,863	8,755 29
Hocking,...	4,008	92,332	215,272	1,919 29
Holmes, ...	9,135	182,439	556,060	6,364 03
Huron, ...	13,346	504,689	1,512,655	15,490 88
Jackson, ...	5,941	57,874	197,932	2,239 69
Jefferson, ..	22,489	230,145	1,855,064	13,149 44
Knox, ...	17,085	313,823	1,252,294	13,329 41
Lawrence,..	5,367	56,862	241,782	2,280 80

Counties.	Population 1830.	Acres of land.	Total amount of taxable property.	Total amount of taxes paid.
Licking, ...	20,869	393,205	2,101,495	17,370 83
Logan, ...	6,440	203,509	519,622	3,925 65
Lorain, ...	5,696	360,863	889,552	10,539 09
Madison, ..	6,190	256,421	600,578	4,643 91
Marion, ...	6,551	168,164	390,602	5,599 78
Medina, ...	7,560	296,257	931,599	10,198 31
Meigs, ...	6,158	229,004	380,172	5,111 58
Mercer	1,110	12,688	54,118	714 30
Miami, ...	12,807	240,093	1,000,748	6,423 09
Monroe, ...	8,768	95,520	280,572	3,666 61
Montgomery,	24,362	267,349	2,293,419	14,649 12
Morgan, ...	11,800	169,135	452,991	4,945 02
Muskingum,	29,334	366,609	2,362,616	18,567 75
Perry, ...	13,970	175,123	729,241	6,116 55
Pickaway,..	16,001	300,969	1,798,665	10,924 76
Pike, ...	6,024	129,153	521,109	4,114 37
Portage, ...	18,826	472,156	2,019,029	17,787 06
Preble, ...	16,291	246,678	1,086,322	7,441 82
Richland, ..	24,008	433,620	1,354,169	15,069 92
Ross, ...	24,068	328,765	2,897,605	17,474 81
Sandusky, .	2,851	95,822	275,992	3,354 64
Scioto, ...	8,740	105,539	963,882	7,926 93
Seneca, ...	6,159	108,758	302,089	3,916 51
Stark, ...	26,588	374,101	1,854,967	16,361 36
Shelby, ...	3,671	66,863	194,468	1,961 26
Trumbull, ..	26,123	556,011	1,807,792	16,635 58
Tuscarawas,	14,298	237,337	902,778	8,955 75
Union, ...	3,192	259,101	380,535	5,193 68
Warren, ...	21,468	243,517	2,143,065	16,247 33
Washington,	11,731	282,498	681,301	7,463 12
Wayne,....	23,333	382,254	1,451,996	14,584 77
Williams and others not incorporated.	1,089	17,797	90,066	1,351 02
Wood,....	1,102	17,981	127,862	1,572 22
Total,	937,903	17,133,481	78,019,526	730,010 75

Statistics for 1836. From the annual report of the auditor of State, it appears there were returned, on the general list for taxation, 17,819,631 acres of land, under the new valuation, made under the law of 1833 '34:

Lands, including buildings, valued at..	$58,166,821
Town lots, including houses, mills, etc.	15,762,594
Horses (262,291), valued at $40 each,	10,491,640
Cattle (455,487), valued at $8 each,..	4,043,896
Merchants' capital and money at interest,	7,262,927
Pleasure carriages (2,603), valued at..	199,518
Total amount of taxable property,	$94,438,016

On the value of taxable property, the following taxes were levied:

State and Canal tax	$142,854 15
County and School tax, ...	396,505 80
Road tax,.....................	66,482 10
Township tax, ...	102,991 65
Corporation, Jail and Bridge tax, ...	51,276 89
Physicians' and Lawyers' tax, ...	3,144 19
School-house tax, ...	1,482 84
Delinquencies of former years,......	13,044 37
Total taxes,	$777,782 07

No returns were made from the counties of Crawford, Hancock, Jefferson and Williams.

Canal Revenues. The total amount of receipts for tolls, for the year ending October 31, 1835, was as follows:

Ohio Canal.

Cleaveland,..	$72,718 72	Newark,....	$20,487 85
Akron,......	6,362 90	Columbus,..	4,605 37
Massillon, ...	13,585 78	Circleville,..	9,651 44
Dover,......	8,096 42	Chillicothe, .	12,134 75
Roscoe,.....	14,555 83	Portsmouth,.	23,118 78
		Total,	$185,317 45

Miama Canal.

Dayton,....	$14,016 75	Hamilton, ...	$3,664 88
Middleton,..	8,747 19	Cincinnati,...	25,803 77
		Total,	$52,232 59

Total tolls received on both canals, ...		$237,550 04
Deduct contingent expenses on Ohio canal,................	$5,836 05	
Do. on Miama canal, ...	2,954 68—	8,790 73
		$228,759 31
Toll received on Lancaster Lat. canal,.		1,062 56
Water rents and sale of State lots, ...		3,700 07
Arrearages paid of tolls received in October, 1834, ...		7,835 26
		$242,357 20

Population of Ohio, at different periods.

In	Population.	From	Increase.
1790, about...	3,000	1790 to 1800, ..	42,365
1800, " ...	45,365	1800 " 1810, ..	185,395
1810, " ...	230,760	1810 " 1820, ..	350,674
1820, " ...	581,434	1820 " 1830, ..	356,469
1830, " ...	937,903	1830 " 1835, ..	437,097
1835, estimated,	1,375,000		

Rivers. The streams which flow into the Ohio river, are the Mahoning, a branch of the Beaver, Little Beaver, Muskingum, Hock-hocking, Scioto, and Little and Great Miami. Those which flow from the northward into lake Erie, are the Maumee, Portage, Sandusky, Huron, Cuyahoga, Grand and Ashtabula. Hence the State is divided into two unequal inclined planes, the longest of which slopes towards the Ohio, and the shortest towards the lake. The head waters of the Muskingum, Scioto and Miami interlock with those of the

Cuyahoga, Sandusky and Maumee, so as to render the construction of canals not only practicable, but comparatively easy. All the large streams are now navigable for boats during the spring season.

Internal Improvements. These consist of canals, rail-roads, turnpike-roads, and the national road, now under the supervision of, and owned by the State. The canaling is managed by a Board of Commissioners. The State canals were projected about 1823; and, considering the youthful character of the State, its want of funds and other circumstances, they are, undoubtedly, the greatest works ever executed in America.

The *Ohio and Erie canal* connects lake Erie with the Ohio river. It commences at Cleaveland, at the mouth of the Cuyahoga, passes along that river and its tributaries, to the summit level; from thence to the waters of the Muskingum, and to the border of Muskingum county; from thence it strikes across the county, past Newark, in Licking county, and strikes the Scioto, down the valley of which it proceeds to its mouth, at Portsmouth. The principal places on the canal are Akron, New Portage, Massillon, Bolivar, New Philadelphia, Coshocton, Newark, Bloomfield, Circleville, Chillicothe, Piketon and Portsmouth. It was commenced on the 4th of July, 1825, and completed in 1832; and, together with the Miami canal to Dayton, cost about $5,500,000, and has greatly enriched the

State and the people. Private property along its line has risen from five to ten fold.

Length of the Ohio and Erie Canal.

	Miles.
Main trunk, from Cleaveland to Portsmouth, . . .	310
Navigable feeder, from main trunk to Columbus, .	11
" " " " " Granville, . .	6
Muskingum side-cut, from the Muskingum river, at Dresden, ...	3
Navigable feeder from the Tuscarawas river, . . .	3
" " " Walhonding river, . . .	1
Total length of Ohio canal and branches,	334

The *Miami canal* commences at Cincinnati; and, passing through the towns of Reading, Hamilton, Middletown, Franklin and Miamisburg, terminates at Dayton, sixty-five miles. It has been navigated from Dayton to the head of Main street, Cincinnati, since the spring of 1829. An extension of the work is now in progress, to be carried along the valleys of St. Mary's and Au Glaise rivers, and unite with the Wabash and Erie canal, at Defiance; distance from Cincinnati about one hundred and ninety miles.

An act passed the Ohio legislature in 1834, for continuing the Wabash and Erie canal (now constructing in Indiana, by that State), from the western boundary of Ohio, to the Maumee bay. Operations have been suspended by the boundary dispute with Michigan.

The *Mahoning and Beaver canal* has already been noticed, under the head of Western Pennsylvania. It is proposed to carry it from Akron, on the Portage summit, along the valley of the Mahoning river, to Newcastle, on the Beaver division of the Pennsylvania canal. Distance in Ohio, seventy-seven miles. The work is in progress.

The *Sandy Creek and Little Beaver canal* is in progress by a chartered company. It commences near the town of Bolivar, on the Ohio and Erie canal, in Tuscarawas county, and passes along near the line of Stark and Carrol counties, to the Little Beaver in Columbiana county, and from thence to the Ohio river.

The *Mad River and Sandusky rail-road* will extend from Dayton on the Miami canal, to Sandusky, through Springfield, Urbanna, Bellefontaine, Upper Sandusky, Tiffin, and down the valley of the Sandusky river, to lake Erie. The route is remarkably favorable for locomotive power. Length 153 miles; estimated cost, $11,000 per mile. The work was commenced in September, 1835.

The *Erie and Ohio rail-road* is intended to be constructed from Ashtabula on the lake, through Warren to Wellsville, on the Ohio river, a distance of ninety miles. Other rail-roads are in contemplation in this State, the most important of which is the *Great Western rail-road*, from Boston, by Worcester, Springfield, and Stockbridge, through New York, by Albany, Utica and Buffalo, along the summit ridge, dividing the northern from the southern waters, through Pennsylvania, Ohio, to intersect the Wabash and Erie canal at Lafayette, in Indiana. From thence provision is already made for it to pass to the eastern boundary of Illinois, from which, a company has been recently chartered to construct it across the State of Illinois by Danville, Shelbyville, Hillsborough to Alton on the Mississippi. It must be some untoward circumstance, that shall prevent this splendid work from being completed the whole length, before 1850.

The project of a rail-road from Cincinnati, to Charleston in South Carolina, has been entered upon with great spirit in the South, and in all the States more directly concerned in the enterprise. It will, undoubtedly, be carried into effect.

The State of Ohio has incorporated a number of turnpike companies, some of which have gone into operation. The first is near the north-eastern corner of the State, from Pierpont, through Monroe and Salem townships to the mouth of Conneaut creek, sixteen miles long. The second is the Trumbull and Ashtabula turnpike, leading from Warren to Ashtabula, forty-eight miles. The third is from the town of Wooster, through Medina to Cleaveland, fifty-one miles. The fourth is from Columbus to Sandusky, 106 miles, now in the course of construction. Another from Cincinnati, through Lebanon and Columbus, to Wooster, has been commenced on the McAdamized plan, but is not completed. A McAdam turnpike from Cincinnati to Chillicothe is in progress. The national road, constructed by the general government, and transferred to the State, passes from Wheeling, through Columbus, to the Indiana line.

Manufactures. The principal factory for woollen goods is at Steubenville. A number of cotten factories are in the towns along the Ohio river. Furnaces for smelting iron ore are in operation in the counties bordering on the Ohio, near the mouth of the Scioto. Glass is manufactured in several towns. Considerable salt is made on the Muskingum below Zanesville, on the Scioto, and on Yellow creek above Steubenville. About half a million of bushels were made in the State, in 1830.

Cincinnati rivals Pittsburgh in the number, variety and extent of its manufacturing operations.

In every town and village through the State, mechanics' shops are established for the manufacture of all articles of ordinary use,

Cities and Towns. To enter upon minute descriptions, or even name all these, would much exceed the bounds of this work.

Cincinnati is the great commercial emporium of the State. It is pleasantly situated on the right or northern bank of the Ohio river, about equidistant from Pittsburgh and its mouth, in north latitude 39° 06′, and west longitude from Washington city 7° 25′.

Directly fronting the city to the south, and on the opposite side of the Ohio river, are the flourishing manufacturing towns of Newport and Covington, which are separated by the Licking river, of Kentucky, which enters the Ohio directly opposite the Cincinnati landing.

The wharf arrangements are the most convenient for lading and unlading goods at all stages of the water, to be found on our western rivers. The town site is beautifully situated on the first and second banks of the river,—the former of which is above ordinary high water, and the latter gently rises sixty or seventy feet higher, and spreads out into a semicircular plain, surrounded with elevated bluffs.

Cincinnati was founded in 1789, but did not grow rapidly till about 1808. The progressive increase of population will appear from the following table:

In	Population.	In	Population.
1810,	2,320	1826,	16,230
1813,	4,000	1830,	26,515
1819,	10,000	1835, estimated,	31,000
1824,	12,016		

Add the adjoining towns of Covington and Newport, whose interests are identified, and the aggregate population will equal thirty-five thousand; and, in all reasonable probability, in 1850, these towns, with Cincinnati, will number one hundred thousand active, educated and enterprising citizens. In 1826, according to the "Picture of Cincinnati," by B. Drake, Esq. and E. D. Mansfield, Esq., the manufacturing industry alone, according to an accurate statistical examination, amounted to $1,800,000. At that time, there were not more than fifteen steam-engines employed in manufactures, in the city. At the close of 1835, there were more than fifty in successful operation, besides four or five in Newport and Covington. "More than one hundred steam-engines, about two hundred and forty cotton-gins, upwards of twenty sugar-mills, and twenty-two steam-boats (many of them of the largest size), have been built or manufactured in Cincinnati during the year 1835."* Hence the productive industry of Cincinnati, Covington and Newport, for 1835, may be estimated at $5,000,000. By a laborious investigation, at the close of 1826, by the same writer, the exports of that year were about $1,000,000 in value. A similar inquiry induced him to place the exports of 1832 at $4,000,000. The estimate for 1835 is $6,000,000.

To enumerate all the public and private edifices deserving notice, would extend this article to too great a length. The court-house, four market-houses, banks, college, Catholic Athenæum, two medical colleges, mechanics' institute, two museums, hospital and lunatics' asylum, the Woodward high

* See a valuable statistical article, by B. Drake, Esq., in the Western Monthly Magazine, for January, 1836, entitled, "*Cincinnati, at the close of* 1835."

school, ten or twelve large edifices for free schools, hotels, and between twenty-five and thirty houses for public worship, some of which are elegant, deserve notice. The type foundry and printing-press manufactory, is one of the most extensive in the United States. Here is machinery, lately invented, for casting printer's types, exceeding, perhaps, any thing in the world. Printing, and the manufacture of books, are extensively carried on in this city. Here are six large bookstores, several binderies, twelve or fifteen printing-offices, from which are issued ten weekly, four tri-weekly, four daily, four monthly and one quarterly publications. Two medical publications, of a highly respectable character, are issued. The Western Monthly Magazine is too well known to need special notice here. The Cincinnati Mirror is a respectable literary periodical. The Family Magazine deserves notice. The Presbyterians, Baptists, Methodists, Roman Catholics, and, perhaps, other sects, have each their weekly paper, respectable in size and character. During four months, in 1831, there were issued from the Cincinnati press, 86,000 volumes, of which 20,300 were original works. In the same period, the periodical press issued 243,200 printed sheets. The business has increased greatly since that time.

The "*College of Professional Teachers,*" is an institution formed at the convention of teachers, held in this city, in October, 1832.

Its objects are to *unite* the professional instructers of youth throughout the Western country in the cause in which they are engaged, and to elevate the character of the profession. Their meetings are held on the first Monday in October annually. Lectures are given, discussions held, reports made, and a respectable volume of transaction published annually. There is no doubt that much good will result to the cause of education in the West, from this annual convocation.

Law School. An institution of this character has been organized, under the management of Hon. J. C. Wright, and other gentlemen of the bar.

Of *Medical Schools* there are two, at the heads of which are gentlemen of high character and attainments in their profession.

The *Mechanics' Institute* is designed for the diffusion of scientific knowledge among the mechanics and citizens generally, by means of popular lectures and mutual instruction. The *Cincinnati Lyceum* was formed for the purpose of useful instruction and entertainment, by means of popular lectures and debates. The *Academic Institute* is designed to aid the cause of education and elevate the profession amongst the teachers in Cincinnati. Its meetings are monthly. The *Athenæum* is an institution under the management of Roman Catholic priests. The college edifice is a splendid and permanent building, of great capacity. The *Woodward High School* was founded by the late William Woodward. The fund yields an income of about $2000 annually. It is conducted by four professors, and has about one hundred and twenty students. The corporation has established a system of free schools, designed to extend the benefits of primary education to all classes, and ten or twelve large edifices have been erected for the purpose. I regret the want of documents to give particulars of this liberal and praiseworthy enterprise, which reflects much honor upon the city and its honorable corporation. In 1833, there were twenty public schools for males and females, and two thousand pupils. Many excellent private schools and seminaries, some of deserved celebrity, are sustained by individual enterprise.

Columbus, the political capital of the State, and nearly in the centre of the State, is a beautiful city, on the east bank of the Scioto river. In 1812, it was covered with a dense forest, when it was selected by the legislature for the permanent seat of government. The public buildings are a state-house, a court-house for the Supreme Court, a building for the public offices, a market-house, &c., all of brick. The State penitentiary is here, for which a new substantial building is constructing, and an Asylum for the deaf and dumb, sustained by legislative aid.

Chillicothe, Cleaveland, Zanesville, Steubenville, Circleville and many others, are large and flourishing towns.

Education. Charters for eight or ten colleges and collegiate institutions have been granted. Congress has granted 92,800 acres of public land to this State, for colleges and academies. One township (23,040 acres), and a very valuable one, has been given to the Miami university, at Oxford. Two townships of land (46,080 acres), though of inferior quality, have been given to the Ohio university. Academies have been established in most of the principal towns. A common school system has been established by the legislature. Each township has been divided into school districts. Taxes are levied to the amount of three fourths of a mill upon the dollar of taxable property in the State, which, with the interest accruing from the different school funds already noticed, are applied towards the expenses of tuition. Five school examiners are appointed in each county, by the court of common pleas, who are to examine teachers. The governer, in his recent message, speaks of the common school system as languishing in proportion to other improvements.

Form of Government. The legislative authority is vested in a Senate and House of Representatives; both of which, collectively, are styled the General Assembly. The members of both branches are chosen by counties, or by districts composed of counties, according to population. The representatives are chosen annually, the senators biennially.

The General Assembly has the sole power of enacting laws; the signature or assent of the governor not being necessary in any case whatever. The judiciary system comprises three grades of courts;—the Supreme Court, courts of common pleas, and Justices' courts. The justices of the peace are chosen triennially, by the people. The executive authority is vested in a governor, who is elected biennially, and must be thirty years of age, and have resided in the State at least four years. He is commander-in-chief of all the militia, and commissions all officers in the State, both civil and military. Each free, white male citizen of the United States, of twenty-one years of age, and a resident of the State one year preceding an election, is entitled to a vote in all elections.

The following shows the professions, occupations, and nativity of the members of the legislature of Ohio, during the present winter (1835–6), and is about a proportionate estimate for other Western States:—

The members of the Ohio legislature, as to their occupations and professions, are:—farmers, fifty-eight; lawyers, seventeen; merchants thirteen; doctors, five; printers, three; surveyors, two; millers, two; masons, two; carpenters, two; painter, one; watch-maker, one; blacksmith, one; house joiner, one.

Their nativity is as follows:—Ohio, seven; Pennsylvania, thirty; Virginia, twenty-two; New England States, seventeen; Maryland, eight; New York, seven; New Jersey, four; Kentucky, three; Delaware, two; North Carolina, one; Ireland, five; England, one; Germany, one.

The youngest member in the Senate, is thirty-three years of age, and the oldest fifty-six. In the House, the youngest twenty-six; oldest sixty-seven. Under the constitution, a senator must be thirty; and a member of the House, twenty-six.

Antiquities. Much has been said about the antiquities of Ohio,—the fortifications, artificial mounds, and military works, supposed to indicate a race of civilized people, as the possessors of the country, anterior to the Indian nations. At Marietta, Circleville, Paint creek, and some other places, are, doubtless, antiquities, that exhibited upon their first discovery, strong marks of a military purpose. I have no doubt, however, that credulity and enthusiasm have greatly exaggerated many appearances in the West, and magnified them into works of vast enterprise and labor. Mounds of earth are found in every country on the globe, of all forms and sizes; and why should they not exist in the western valley? Mr. Flint states that he has seen a horse-shoe dug up at the depth of thirty-five feet below the surface, with nails in it, and much eroded by rust. He mentions also a sword, which is *said* to be preserved as a curiosity, but which he had not seen, found enclosed in the wood of the roots of a tree, which could not have been less than five hundred years old! Those who delight especially in the marvellous, may consult the "Description of the Antiquities discovered in the State of Ohio and other Western States, by Caleb Atwater, Esq."

History. The first permanent settlement of Ohio, was made at Marietta, April 7, 1788, by forty-seven persons from Massachusetts, Rhode Island, and Connecticut. This was the nucleus around which has grown up the populous State of Ohio. Amongst the most active promoters of this colony, were those called then "The Ohio Company." The next settlement was that of Symmes' purchase, made at Columbia, six miles above Cincinnati, in November, 1789, by Major Stites and twenty-five others, under the direction of Judge Symmes. A colony of French emigrants settled at Gallipolis, in 1791. In 1796, settlements were made by New England emigrants at Cleaveland and Conneaut, on the southern shore of lake Erie. The intermediate country gradually filled up by emigration from various parts of the United States. Some slight diversity exists, in different sections of the State, in manners, customs and feelings, amongst the people, in accordance with the States or countries from which they or their fathers emigrated. These shades of character will become blended, and the next generation will be *Ohions*, or, to use their own native cognomen, *Buckeyes.*

The first territorial legislature convened at Cincinnati, in September, 1790. The governor having exercised his right of *veto* in relation to the removal of a county seat, an unhappy collision followed, and, upon framing the State constitution, in November, 1802, the convention prevented the governor of the State from ever exercising the *negative* power upon acts of the legislature.

Date of Organization of some of the oldest Counties.

Washington,	July 27, 1788
Hamilton,	Jan. 2, 1790
Adams,	July 10, 1797
Jefferson,	July 29, 1797
Ross,	Aug. 20, 1798
Trumbull,	July 18, 1800
Clermont,	Dec. 6, 1800
Belmont,	Sept. 7, 1801

These were all organized under the territorial government.

CHAPTER XI.

INDIANA.

Boundaries and Extent—Counties—Population at different Periods—Face of the Country—Sketch of each County—Form of Government—Finances—Internal Improvements—Manufactures—Education—History—General Remarks.

LENGTH two hundred and forty, breadth one hundred and fifty miles; between 37° 48′ north latitude, and 7° 45′ and 11° west longitude; bounded north by the State of Michigan and lake Michigan, east by Ohio, south by the Ohio river, which separates it from Kentucky, and west by Illinois. It contains about 37,000 square miles, equal to 23,680,000 acres.

It is naturally subdivided into the hilly portion, bordering on the Ohio; the level, timbered portion, extending across the middle of the State; the Wabash country, on that river; and the northern portion, bordering on the State of Michigan and the lake. The two last portions include nearly all the prairie country.

For civil purposes, this State has been divided into counties, and those subdivided into townships.

Table of Counties, Seats of Justice, &c.

Counties.	Date of Formation.	Square miles.	Population in 1830.	Seats of Justice.
Allen, ...	1823	720	1,000	Fort Wayne.
Bartholomew,...	1821	588	5,800	Columbus.
Boon, ...	1830	400	622	Lebanon.
Carroll, ...	1828	450	1,614	Delphi.
Cass, ...	1829	460	1,154	Logansport.
Clark, ...	1802	400	10,719	Charlestown.
Clay, ...	1825	360	1,616	Bowling Green.
Clinton,........	1830	450	1,423	Frankfort.
Crawford, ...	1818	350	3,184	Fredonia.
Daviess, ...	1816	460	4,512	Washington.
Dearborn, ...	1802	448	14,573	Lawrenceville.
Decatur, ...	1821	400	5,854	Greensburgh.
Delaware, ...	1827	440	2,372	Muncytown.
Dubois,........	1817	420	1,774	Jasper.
Elkhart, ...	1830	576	935	Goshen.
Fayette, ...	1818	200	9,112	Connersville.
Floyd, ...	1819	200	6,363	New Albany.
Fountain, ...	1825	400	7,644	Covington.
Franklin, ...	1810	400	10,199	Brookville.
Gibson, ...	1813	450	5,417	Princeton.
Grant, ...	1831	415	——	Marion.
Greene,........	1821	540	4,250	Bloomfield.
Hamilton,......	1823	400	1,705	Noblesville.
Hancock, ...	1828	340	1,569	Greenfield.
Harrison,..	1808	470	10,288	Corydon.
Hendricks, ...	1823	420	3,967	Danville.
Henry, ...	1821	440	6,498	Newcastle.
Huntington,.....	1832	400	——	
Jackson, ...	1815	500	4,894	Brownstown.
Jefferson, ...	1809	400	11,465	Madison.
Jennings,.......	1816	400	3,950	Vernon.
Johnson, ...	1822	300	4,130	Franklin
Knox, ...	1802	540	6,557	Vincennes.
La Porte, ...	1832	420	——	La Porte.

Counties.	Date of Formation.	Square miles.	Population in 1830.	Seats of Justice.
Lagrange, ...	1832	380	——	Mongoquinon.
Lawrence,.......	1818	460	9,237	Bedford.
Madison,........	1823	420	2,442	Andersontown.
Marion,.........	1821	440	7,181	INDIANOPOLIS.
Martin, ...	1818	340	2,010	Mount Pleasant.
Miami, ...	1832	330	——	Miamisport.
Monroe, ...	1818	560	6,578	Bloomington.
Montgomery,....	1822	500	7,376	Crawfordsville.
Morgan, ...	1821	530	5,579	Martinsville.
Orange,........	1815	378	7,909	Paoli.
Owen, ...	1818	380	4,060	Spencer.
Parke, ...	1821	450	7,534	Rockville.
Perry, ...	1814	400	3,378	Rome.
Pike, ...	1816	430	2,464	Petersburgh.
Posey, ...	1814	500	6,883	Mount Vernon.
Putnam, ...	1821	490	8,195	Greencastle.
Randolph, ...	1818	440	3,912	Winchester.
Ripley, ...	1818	400	3,957	Versailles.
Rush, ...	1821	400	9,918	Rushville.
Scott,.........	1817	200	3,097	Lexington.
Shelby, ...	1821	430	6,294	Shelbyville.
Spencer, ...	1818	400	3,187	Rockport.
St. Joseph, ...	1830	740	287	South Bend.
Sullivan, ...	1816	430	4,696	Merom.
Switzerland, ...	1814	300	7,111	Vevay.
Tippecanoe, ...	1826	500	7,161	Lafayette.
Union, ...	1821	224	7,957	Liberty.
Vanderburgh, ...	1818	225	2,610	Evansville.
Vermilion,......	1823	280	5,706	Newport.
Vigo, ...	1818	400	5,737	Terre Haute.
Wabash, ...	1832	380	——	
Warren, ...	1828	350	2,854	Williamsport.
Warrick,.......	1813	412	2,973	Boonville.
Washington,.....	1813	550	13,072	Salem.
Wayne, ...	1810	420	23,344	Centreville.

The total population is 1830 was 341,682; the estimated population in the message of Gov. Noble, to the legislature, December, 1835, was 600,000.

The counties in which the population has not been given in the foregoing table, have been formed since 1830. Probably other new counties, along the waters of the Wabash and Kankakee, have been formed recently, of which no intelligence has been had by the author. The counties in the northern portion of the State have increased the most in population since 1830.

For electing representatives to Congress, the State is divided into seven electoral districts: for judicial purposes, it is divided into eight circuits, in each of which there is a circuit judge, who, together with two associates in each county, holds the circuit courts.

Population at Different Periods.

In	Population.	From	Increase.
1800 (excluding Illinois), ...	2,641	1800 to 1810, . .	21,879
1810, ...	24,520	1810 " 1820, . .	122,658
1820, ...	147,178	1820 " 1825, . .	74,822
1825, ...	222,000	1825 " 1830, . .	119,582
1830, ...	341,582	1830 " 1835, . .	258,418
1835, estimated,	600,000		

In 1825, the number of voters was 36,977, and the number of paupers, 217.

Face of the Country, &c. The counties bordering on the Ohio river are hilly; sometimes abrupt, precipitous, stony, and occasionally degenerating into knobs and ravines. Commencing at the mouth of White river, on the Wabash, and following up that stream, on its east fork, and thence along the Muscatatack, through Jennings and Ripley counties, to Lawrenceville, and you leave the rough and hilly portion of Indiana to the right. Much of the country we have denominated hilly is rich, fertile land, even to the summit of the hills. On all the streams are strips of rich alluvion of exhaustless fertility. The interior, on the two White rivers and tributaries, is moderately undulating, tolerably rich soil, and much of it heavily timbered with oaks of various species, poplar, beech, sugar-tree, walnut, hickory, elm and other varieties common to the West. There is much level, table land, between the streams. Along the Wabash, below Terre Haute, is an undulating surface, diversified with forest and prairie, having a soil of middling quality, interspersed with some very rich tracts. Above Terre Haute, along the Wabash and its tributaries, the land in general is first rate; a large proportion forest, interspersed with beautiful prairies. The timber consists of oak of various species, poplar, ash, walnut, cherry, sugar-tree, buckeye, hickory, beech, sassafras, linden, honey-locust, with some cottonwood, sycamore, hackberry and mulberry on the bottom lands. The undergrowth is spicebush, hazel, plum, crab-apple, hawthorn and vines. Along the northern part of the State are extensive prairies and tracts of barrens, with groves of various kinds of timber, and skirts of burr-oak. Towards lake Michigan, and along the Kankakee and St. Joseph rivers, are lakes, swamps and marshes.

Rivers. The Ohio meanders along the south-eastern and southern parts of the State for 350 miles. The east and west forks of White river, and their tributaries, water the interior counties for 100 miles in extent. They are both navigable streams for flat-boats during the spring and autumn floods. The Wabash river has several heads, which interlock with the waters of the St. Joseph and St. Mary's, which form the Maumee of lake Erie. It runs a south-westwardly course across the State, to Warren county,—thence southwardly to Vigo county, where it becomes the boundary between Indiana and Illinois, along which it meanders to the Ohio, which it enters twelve miles above Shawneetown. The St. Joseph of lake Michigan, already noticed under the State of Michigan, makes a curve into Elkhart and St. Joseph counties, forming what is called the *South Bend.* The Kankakee, which is the longest branch of Illinois river, rises in Indiana, near the South Bend. Some of its head waters interlock with those of Tippecanoe, a prominent tributary of the Wabash.

Sketch of each County. The following sketch of each county, its streams, surface, soil and minerals, has been made and collated with much labor, from an excellent Gazetteer of this State, published in 1833, by Douglass & Maguire, of Indianapolis, from personal observation of many of the older counties, and from an extensive correspondence.

Allen.—Streams; Saint Joseph's and Saint Mary's, which form the Maumee of lake Erie (navigable for small keel-boats), and numerous creeks; generally heavily timbered; soil clay, sandy on the rivers.

Bartholemew.—Streams; Driftwood, Clifty, Flat Rock, and Salt Creeks,—all mill streams. Surface, level; soil, a rich loam, mixed with sand and gravel; the western part hilly, with clay soil. Minerals; limestone, coal, iron ore, red ochre.

Boon.—Watered by the tributaries of Racoon and Sugar Creeks. Surface, level,—soil rich.

Carroll.—Streams; Wabash river, Deer, Rock, and branches of Wildcat creeks. Considerable timber,—some prairies, of which Deer prairie is the largest and most beautiful. Considerable quantities of limestone on the surface; a remarkable spring near Delphi,—the water reddish.

Cass.—Streams are Wabash and Eel rivers, which unite at Logansport,—the head of steam-boat navigation of the Wabash. Surface, generally level, rolling towards the rivers with abrupt bluffs; soil, near the rivers, a mixture of loam and sand; at a distance from them, flat and clayey. Large proportion forest land,—some prairies.

Clark.—Silver and Fourteen Mile creeks furnish excellent mill sites. Ohio river on the south. Surface, rolling and hilly; soil, loam, mixed with sand. Minerals; limestone, gypsum, water lime, marble, salt, iron ore, copperas, alum.

Clay.—Eel river and tributaries. Surface moderately undulating; soil various, chiefly clay and loam, and a mixture of sand, in places; timber predominates,—some prairies.

Clinton.—Watered by the South, Middle, and Kilmore's Forks of Wildcat creek. Surface, moderately undulating, or level: Twelve Mile prairie extends from south-west to north-east twelve miles, and is three fourths of a mile wide. The remainder timbered land. Soil, a rich sandy loam, and exceedingly fertile.

Crawford.—Waters; the Ohio and Blue rivers,—plenty of water power, and excellent springs. Surface, hilly and broken; in places, tolerably productive; in others, soil thin and rocky. A timbered region, and abundance of limestone.

Daviess.—Streams; Forks of White river, with its tributaries, Smother's, Prairie, Veal, Aikman's and Sugar creeks. Level bottoms on the rivers—sometimes inundated; undulating on the high grounds. Soil on the West Fork, sandy; much timber,—an extensive tract of sugar tree; some prairies. The county destitute of rock near the surface; plenty of lime and sandstone in the bed of West Fork of White river, at the rapids. Plenty of coal.

Dearborn.—Watered by the Great Miami, Whitewater, Laughery, Hogan's and Tanner's creeks. Surface hilly and broken, with rich, level, bottom lands, on the Miami. Soil, one fourth first rate, one fourth second rate, —remainder inferior. A timbered region.

Decatur.—Flat Rock, Clifty, and Sand creeks, are all good mill streams. Surface, generally level,—some parts undulating; soil, loam, with a substratum of clay; well adapted to grain,—timbered. Minerals; limestone, some iron ore and coal.

Delaware.—Streams; Missisinawa, and West Fork of White river; surface tolerably level; soil, loam, mixed with sand. Minerals; some limestone, and granite bowlders scattered over the surface.

Dubois.—Streams; East Fork of White river, Patoka and Anderson creeks. Surface rolling,—some parts hilly and broken,—some level tracts; soil, rich and sandy loam near the streams. Minerals; sandrock and coal.

Elkhart.—Watered by St. Joseph of lake Michigan, Elkhart and tributaries. Surface, generally level,—a portion undulating; soil various, but generally rich; forest and prairie, both wet and dry.

Fayette.—Watered by the West Fork of Whitewater, and a small lake in the north. Surface, undulating; soil, on the high ground, clayey, and a mixture of sand,—on the bottom lands, a rich, sandy loam. Limestone found in masses and quarries.

Floyd.—Watered by the Ohio river, Silver creek, and some head branches of Big and Little Indian creeks. Surface various,—a range of knobs,—east of these knobs, it is gently undulating; soil inferior. Minerals; shale, soft sandstone, limestone, freestone, iron ore, and some traces of coal. A boiling spring, from which is emitted an inflammable gas.

Fountain.—Watered by the Wabash river, and Coal and Shawnee creeks, with numerous mill sites. Surface, gently undulating; soil, a black loam, mixed with sand and very rich. Minerals; coal, and some sandstone.

Franklin.—Watered by the East and West Forks of Whitewater. Surface, on the eastern part level,—western, rolling; soil, in the central and northern parts, a black loam,—in the south-west, thin and clayey.

Gibson.—Watered by the Wabash, White, and Patoka rivers. Surface, rolling and timbered; soil, generally a sandy loam, and productive.

Grant.—Watered by the Missisinawa and tributaries. Surface level,—generally heavily timbered; soil, clay and loam on the table lands,—sandy on the river bottoms.

Green.—Watered by White and Eel rivers, and Richland creek; soil, on the rivers, a rich loam,—on the bluffs, sandy,—east side, hilly, —west side, level. White river is navigable. Minerals; lime and sandstone, coal, and some iron ore.

Hamilton.—The streams are White river, and Cicero, Coal, Stoney, and Fall creeks. Generally forest,—some few prairies; soil, in places, clay,—more generally, a sandy loam. Minerals; lime, and some soft sandrock.

Hancock.—Watered by Blue river, Sugar and Brandywine creeks, with excellent mill sites, and well supplied with springs. Surface, either level or gently undulating; soil, a rich loam, mixed with sand,—heavily timbered.

Harrison.—Watered by Big and Little Indian, and Buck creeks, and Blue river. Surface various,—some parts hilly and broken,—some parts undulating,—some parts level; soil, in the low grounds, a rich loam,—on the high grounds, calcareous and gravelly. A large tract of "barrens" in the west. Minerals; a quarry and several caves of black flint, salt licks, limestone.

Hendricks.—The waters are White Lick, and branches of Eel river, with good mill sites. Surface, gently rolling, and timbered with the varieties of the Wabash country; soil, a mixture of clay, loam and sand.

Henry.—Watered by Blue river, Flat Rock and Fall creeks. Surface, in some places, broken,—in most parts, level; soil, a mixture of sand with loam and clay. Plenty of springs and mill sites. Mostly timbered, but several tracts of prairie.

Huntington.—The streams are Salamania, Little river, and Wabash. Surface, on the rivers, level,—back, gently undulating; soil, loam and clay, with a slight mixture of sand. Several tracts of prairie, but generally forest land.

Jackson.—Watered by Indian, Driftwood, White, Muscatatack, and Gum creeks. Surface, rolling and in places hilly; soil, clay and loam, mixed with sand. In the forks of the creeks, sand predominates. On the west and north-west, inclined to clay.

Jefferson.—Watered by the Ohio river, Indian-Kentucky and Big creeks. Surface various; along the river and creeks, low alluvion; soil, loam mixed with sand. The bottoms are bounded by precipitous bluffs, with towering cliffs of limestone. The table lands are undulating, and the soil inclined to clay. Timber various. Abounds with limestone, masses of freestone, and scattered granite bowlders.

Johnson.—Watered on the eastern side by Blue river, and Sugar and Young's creeks,—on the western side by Indian, Crooked, and Stott's creeks. Surface, gently undulating; soil, a rich, black, sandy loam; timbered. Minerals; masses of freestone, and scattered granite bowlders.

Jennings.—Watered by Graham's Fork, and the North Fork of the Muscatatack. Surface, in some parts level, some parts very hilly; soil, calcareous, rich and productive; timber of all varieties; abounds with limestone.

Knox.—The Wabash on the west side,—White river south,—the West Fork of White river east,—and Maria and Duchain creeks, interior. Surface undulating; soil, somewhat various,—a rich loam in places,—sandy in other places;—some tracts of prairie, but timber predominates.

Lagrange.—Watered by Pigeon and Crooked rivers. Surface, gently rolling; northern part extensive prairies; southern portion chiefly forest; soil, loam and sand.

La Porte.—Watered by the Kankakee, Galena, and Trail creek, at the mouth of which is Michigan city and a harbor for lake Michigan commerce. Surface, gently undulating; abounds with large, rich prairies, with groves of timber, and lakes of clear water interspersed; soil, a sandy loam, rich and productive.

Lawrence.—Watered by Salt, Indian, Guthrie's, Beaver, and Leatherwood creeks, and excellent springs. Surface, generally hilly,—some level lands;—soil, on the water courses, sandy,—back from the streams, loam and clay. Abounds with limestone.

Madison.—The West Fork of White river is navigable. The other streams are Killbuck, Pipe, Lick and Fall creeks. Surface, generally level, with some broken land near the streams; timbered, with a wet prairie, seven miles long and three fourths of a mile wide; soil, sand, mixed with clay and loam,—productive. Minerals; lime and freestone, marble that polishes well, and some traces of iron ore.

Marion.—West Fork of White river passes through it, on which is situated INDIANOPOLIS, the capital of the State. Fall creek is an excellent mill stream. Surface, chiefly level forest land; soil, a deep black loam, with a mixture of sand. Large granite bowlders are scattered over the surface.

Martin.—The East Fork of White river passes through it, and receives Lost river from the left, and Indian and Flint creeks from the right. Surface, on the east side of White river, broken and hilly; soil, clay and loam; on the west side, level, or gently undulating, with portions of barrens and prairie land; soil, clay and loam, mixed with sand. Minerals; coal in large quantities, lime, sand and freestone.

Miami.—The Wabash and Eel rivers pass through it, and the Missisinawa comes from the east, and enters the Wabash about the centre of the county. The Wabash and Erie canal passes through it. Surface, gently undulating and beautiful,—chiefly forest, and interspersed with small prairies; soil, the richest in the State, of loam, clay and sand intermixed.

Monroe.—Streams; Salt, Clear, Indian, Racoon, Richland, and Bean-blossom creeks, —pure springs. Surface, hilly and undulating; soil, second rate. Minerals; limestone rock, salt licks, with manufactories of salt.

Montgomery.—The heads of Shawnee and Coal creeks in the north-west,—Sugar creek in the centre,—and Big Racoon on the south-eastern part. Surface, gently undulating; the northern portion prairie, interspersed with groves, with a rich soil of black loam, mixed with sand,—the middle and southern portions timbered. Excellent quarries of rock in the middle,—granite bowlders in the northern parts.

Morgan.—White river, which is navigable. The mill streams are White Lick, Sycamore, Highland, and Lamb's creeks on the west side, and Crooked, Stott's, Clear, and Indian creeks on the east side. Surface, generally rolling,—some parts hilly; soil, calcareous and clayey,—on the bottoms, a rich sandy loam. Minerals; limestone, and some iron ore.

Orange.—Streams; Lost river, French Lick, and Patoka. Surface, hilly and broken,—limestone rock,—springs of water, of which Half-moon and French Lick are curiosities. On the alluvial bottoms, the soil is loamy,—on the hills, calcareous, and inclined to clay. Excellent stones for grit, equal to the Turkey oil stones are found in this county.

Owen.—Watered by the West Fork of White river, with its tributaries, Racoon, Indian, Mill, Rattlesnake, and Fish creeks. The falls of Eel river furnish the best water power in the State. Surface rolling; soil, in some places a dark loam,—in others clayey and calcareous. Minerals; immense bodies of lime rock, and some iron ore.

Parke.—Watered by the Big and Little Racoon, and Sugar creeks (with excellent mill sites), all of which enter the Wabash on its western side. Surface, generally level,—some beautiful prairies, but mostly forest land; soil, a loam, mixed with sand, and rich. Minerals; lime and sandstone, coal and iron ore.

Perry.—Watered by the Ohio river, with Anderson's, Bear, Poison, and Oil creeks interior. Some level land, with a rich, sandy loam, on the streams,—all the high lands very broken; hilly, with a clayey, sterile soil. Minerals; immense bodies of limestone, grindstone quarries, iron ore and coal.

Pike—has White river on the north, and Patoka creek through the centre. Surface all forest land and undulating; soil, eastern part clay and sand,—western, a rich, dark loam, mixed with sand,—some swampy land. Minerals; limestone and coal.

Posey.—In the forks of the Ohio and Wabash, with Big, Mill, and McFadden's creeks interior, and good springs. Surface, rolling, and all forest land; soil, a sandy loam, and produces well. Minerals; sand, and limestone, and coal.

Putnam—has Racoon creek, and Eel river, with abundant water privileges, and fine springs. Surface, gently undulating; soil, in places calcareous and clayey,—in other places a rich loam; limestone.

Randolph.—Water courses; the West Fork of White river and Missisinawa and their tributaries, which furnish good mill sites. Surface, either level or gently undulating; soil, a rich loam,—in some places marshy; a small quantity of limestone, with granite bowlders.

Ripley.—Watered by Laughery and Graham's creek. Surface, level, forest land; soil clay,—in some parts inclines to sand,—with limestone abundant.

Rush.—The streams are Big and Little Blue river, Big and Little Flat Rock, with excellent water power. Surface, moderately rolling, and heavily timbered; soil, loam on clay, with a slight mixture of sand.

Scott.—Watered by tributaries of the Muscatatack. Surface rolling,—some flat lands, inclining to marsh; soil, clay; minerals, limestone, iron ore, salt, sulphur and copperas.

Shelby.—Watered by Big and Little Blue river, Brandywine and Sugar creeks, with good mill sites,—all heads of the east fork of White river. Surface, generally level, with forest land; soil, clay, mixed with loam.

Spencer.—Ohio river, Anderson's, Little Pigeon and Sandy creeks. Surface tolerably level, and forest land; soil, clay, mixed with loam: minerals; coal, lime, and sandrock.

St Joseph.—St. Joseph's river, Kankakee and Bobango, with some small creeks; extensive marshes on the Kankakee, and near the South Bend of the St. Joseph. These marshes are of vegetable formation. Surface, in some parts level, in others, gently undulating; soil, a loam, in some places sand. The north-west part chiefly prairies and barrens, including the large and fertile prairies of Portage and Terre Coupee. The north-eastern, barrens; the south-eastern, forest. Minerals, granite bowlders and bog-iron ore.

Sullivan—has the Wabash river on its western side, and Turman's, Busseron and Turtle creeks, interior. Surface, rolling;—some prairies, but generally forest land,—some poor barrens; soil, loam and sand;—lime, sandrock and coal.

Switzerland.—The Ohio east and south,—Indian, Plum, Bryant's, Turtle and Grant's creeks, interior. Surface, various,—bottom lands level and rich,—then a range of precipitous bluffs, with cliffs of limestone,—the table land rolling, with a calcareous and clayey soil. At Vevay are extensive vineyards.

Tippecanoe.—Watered by the Wabash river, and Wildcat, Wea, Burnett's and Mill Branch creeks. The Wabash affords navigation, and the other streams excellent mill sites. Surface gently undulating, with extensive level tracts, and consists of one half prairie, one eighth barrens, and the remainder heavy forest land. The prairie soil is a rich, black loam,—the barrens cold, wet clay,—the forest a very rich loam and sand.

Union.—Streams; the East Fork of White river and its tributaries, Hanna's, Richland

and Silver creeks, all of which furnish excellent mill sites. Surface, moderately rolling; soil, a dark loam.

Vanderburgh.—Watered by the Ohio and Great Pigeon creek. Surface, high, dry, rolling land, with good timber, and well watered; soil, clay and sand, of inferior quality. Minerals; lime and sandstone, salines and a mineral spring.

Vermilion.—A long, narrow county, between the Wabash river and the State of Illinois. The streams are Wabash, Big and Little Vermilion, and their tributaries. Surface high, rolling land, with abrupt bluffs near the streams; a good proportion of prairie and timber; soil, rich, sandy loam, and very productive. Minerals; freestone, limestone, and large coal banks.

Vigo.—The Wabash passes through it,—navigable. The mill streams are Prairie, Honey, Otter and Sugar creeks, but their waters fail in a dry season. Surface level, or gently undulating, with forest and prairies; soil, first rate, rich loam and sand. Minerals; gray limestone, freestone, and inexhaustible beds of coal.

Wabash.—The Wabash river, and Wabash and Erie canal, pass through it, as does the Missisinawa, Eel, Bluegrass and Salamania. Surface, wide, rich bottoms on the streams; bluffs and ravines adjoining; table lands further back, either dry and rolling, or flat and wet, and abounding with willow swamps.

Limestone rock abundant, and many excellent springs of pure water.

Warren.—The Wabash on the south-east border for thirty miles, and navigated by steam-boats; interior streams, Rock, Redwood, and Big and Little Pine creeks, all of which afford good mill sites; some pine and cedar timber. Surface generally level, with broken land on the bluffs of creeks; some forest, but the largest proportion prairie; soil, a rich and very fertile loam. Minerals; lime and excellent freestone for building purposes, coal, iron, lead and copper, with several old "diggings" and furnaces, where copper and lead ore have been smelted in early times.

Warrick.—Watered by the Ohio river, Big and Little Pigeon and Cypress. Surface, rolling and hilly; soil, a sandy loam, on clay. Minerals; quarries of freestone, some limestone, and inexhaustible beds of coal.

Washington.—Streams; Muscatatack, on the north, Rush, Twin, Highland, Delany's, Elk, Bear and Sinking creeks, and the heads of Blue and Lost rivers, and mill sites. Surface, diversified, from gentle undulations, to lofty and precipitous hills; soil, in part, second rate, with much of inferior quality: substratum of limestone; caves, hollows and sink-holes.

Wayne.—Streams; East and West Forks of Whitewater, with excellent water-power for machinery. Surface, moderately hilly, heavy forest land; soil, a rich loam; substratum, clay. Minerals; generally limestone, excellent for building.

Form of Government. This differs very little from that of Ohio. The constitution provides that an enumeration be made every five years of all free white male inhabitants, above the age of twenty-one years; and the representation of both houses of the General Assembly is apportioned by such enumeration, in such ratio that the number of representatives shall never be less than thirty-six, nor exceed one hundred, and the number of senators not exceeding one half nor less than one third the number of representatives. Every free white male citizen, twenty-one years of age, who has resided in the State one year, is entitled to vote; "except such as shall be enlisted in the army of the United States, or their allies." Elections are held annually, by ballot, on the first Monday in August. The governor, lieutenant governor and senators, hold their offices for three years. The judiciary is vested in a Supreme Court, in circuit courts, probate courts, and justices of the peace. The Supreme Court consists of three judges, who are appointed by the governor, with the advice and consent of the Senate, for the term of seven years, and have appellate jurisdiction. The circuit courts consist of a presiding judge in each judicial circuit, elected by joint ballot of both houses of the General Assembly, and two associate judges in each county, elected by the qualified voters, in their respective counties, for a like term. The probate courts consist of one judge for each county, who is elected by the voters, for the same term. Justices of the peace are elected in each township, for the term of five years, and have jurisdiction, in criminal cases, throughout the county, but in all civil cases, throughout the township.

Finances. The Indiana Gazetteer, of 1833, estimates that the revenue for State purposes, amounted to about $35,000 annually; and, for county purposes, to about half that sum. The aggregate receipts for 1835, according to the governor's message of Dec. 1835, amounted to $107,714; expenditures for the same time, $103,901. Sales of canal lands for the same period, $175,740. The canal commissioners have borrowed $605,257, for canal purposes, on a part of which they obtained two per cent. premium, and, on another part, as high as seven per cent.; and have also borrowed $450,000 bank capital, for which they received four and a half per cent. premium. Three per cent. on all sales of United States lands within the State is paid by the general government into the State treasury, to be expended in making roads. The receipts from this source, in 1835, amounted to $24,398. Sales and rents of saline lands, produced an income of $4,636. The proceeds of certain lands, donated by the general government towards the construction of a road from the Ohio river to lake Michigan, amounted to $33,030.

Internal Improvements. This State has entered with great spirit upon a system of internal improvements. It consists of canaling, improving river navigation, rail-roads, and common turnpike-roads.

Wabash and Erie Canal. This work will extend from Lafayette, on the Wabash river, up the valley of that stream, to the Maumee, and to the boundary of Ohio, a distance of one hundred and five miles. The cost of construction has been estimated at $1,081,970, and lands to the amount of 355,200 acres, have been appropriated by the general government, the proceeds of which will be sufficient to complete the canal to Fort Wayne. The middle division, thirty-two miles, was completed in July, 1835, and the remainder is in active progress. Its whole distance, through a part of Ohio, to Maumee bay, at the west end of lake Erie, will be one hundred and eighty-seven miles.

The *Whitewater Canal*, seventy-six miles in length, along the western branch of Whitewater, is intended to pass through Connersville, Brookville, Somerset and other towns, to Lawrenceburgh, on the Ohio river.

Provision is made to improve the navigation of the Wabash river, in conjunction with Illinois, where it constitutes the boundary line; and by this State alone, further up.

Synopsis of canals surveyed by order of the Indiana legislature, in 1835.—Lafayette and Terre Haute division of the Wabash and Erie canal; length, ninety miles; total cost, $1,067,914 70; per mile, $11,865 79.

Central Canal, north of Indianopolis: total length from Indianopolis, by the way of Andersontown, Pipe creek summit, to the Wabash and Erie canal, at Wabash town, one hundred and three miles, thirty-four chains; total cost, $1,992,224 54; per mile, $17,106 51:—length, by the way of Pipe creek summit, to Peru, near the mouth of the Missisinawa, one hundred and fourteen miles, forty-six chains; total cost, $1,897,797 19; per mile, $14,871 85:—length, by the way of Pipe creek summit (including lateral canal to Muncytown), to Wabash town, one hundred and twenty-four miles, fifty-one chains; total cost, $2,103,153 61; per mile, $15,873 83:—length, by the way of Pipe creek summit (including lateral canal to Muncytown), to Peru, one hundred and eighty-five miles, sixty-three chains; total cost, $2,008,726 26; per mile, $14,793 12.—Total length, from Indianopolis, by the way of Muncytown, to the Wabash and Erie canal, at Peru, one hundred and thirty-one miles, forty-one chains; total cost, $2,058,929 41; per mile, $14,549 71. Central canal, south of Indianopolis: total length, from Indianopolis to Evansville, one hundred and eighty-eight miles; total cost, $2,642,285 92; per mile, $14,054 71. Route down the valley of Main Pigeon,—length, one hundred and ninety-four miles; total cost, $2,400,957 70; per mile, $12,376 02.

Terre Haute and Eel river Canal, which forms a connexion between the Wabash and

Erie canal and White river or Central canal: total length, forty miles and a half; total cost, $629,631 65; which, including a feeder, is $13,540 46 per mile.

Wabash and Erie Canal, eastern division, (east of Fort Wayne); upper line,—length, nineteen miles, thirty chains; total cost, $154,113 13; per mile, $7,952 17: lower line, —total length, twenty miles, seventy-six and a half chains; total cost, $254,817 52; per mile, $11,159 04.

The following are the works provided for in the bill, and the sums appropriated for them:

1. The Whitewater canal, including a lateral canal or rail-road, to connect said canal with the Central or White river canal, ...	$1,400,000
2. Central or White river canal, ...	3,500,000
3. Extension of the Wabash and Erie canal, ...	1,300,000
4. Madison and Lafayette rail-road, . . .	1,300,000
5. M'Adamized turnpike-road from New Albany to Vincennes, ...	1,150,000
6. Turnpike or rail-road, from New Albany to Crawfordsville, ...	1,300,000
7. Removing obstructions in the Wabash,	50,000
	$10,000,000
8. The bill gives the credit of the State to the Lawrenceburgh and Indianapolis rail-road company, for the sum of	$500,000

Rail-Roads, from Evansville, on the Ohio, to Lafayette, on the Wabash, one hundred and seventy-five miles;—from Lafayette to Michigan city, ninety miles, forming a line from the Ohio river to lake Michigan, two hundred and sixty-five miles in length;—from Madison, on the Ohio, to Indianopolis, the seat of government, eighty-five miles; and several others, were projected three years since. But at the session of the legislature of 1835–6, a bill was passed to borrow, in such instalments as should be needed, *ten millions* of dollars; and a system of internal improvements, including canals, rail-roads, and the improvement of river navigation, was marked out. In a few years, this State will be prominent in this species of enterprise.

Manufactures. Besides the household manufacture of cotton and flannels, common to the western people, at Vincennes, and probably other towns, machinery is employed in several establishments. It will be seen from the sketch of each county, already given, that in most parts of the State there is a supply of water power for manufacturing purposes. Both water and steam power, saw and grist-mills, are already in operation in various parts of the State.

Education. The same provision, of one section of land in each township, or a thirty-sixth part of the public lands, has been made for the encouragement of common schools, as in other Western States. A law has been enacted providing for common schools, and the public mind has become in a measure awakened to the subject of education. Some most extravagant and exaggerated statements have been made, relative to an incredible number of children in this State, "who have no means of education." As in all new countries, the first class of emigrants, having to provide for their more immediate wants, have not done so much as is desirable to promote common school education; but we have no idea they will slumber on that subject, while they are wide awake to the physical wants and resources of the country. Academies have been established in several counties, and a college at Bloomington, from the encouragement of State funds; and other institutions are rising up, of which the Hanover institution, near the Ohio river, and Wabash college, at Crawfordsville, promise to be conspicuous.

History. This country was first explored by adventurers from Canada, with a view to the Indian trade, towards the close of the seventeenth century; and the place where Vincennes now stands is said to have been thus early occupied as a trading post. A company of French, from Canada, made a settlement here, in 1735. The country, in common with the Western Valley, was claimed by France, until it was ceded to Great Britain, at the treaty of peace, in 1763, under whose jurisdiction it remained, until subdued by the American arms, under the intrepid Gen. G. R. Clark, and his gallant band, in 1779. A territorial government was organized by Congress, in 1787, including all the country northwest of the Ohio river, which was then called the North-Western Territory. In 1802, when the State of Ohio was organized, all that part of the Territory lying west of a line due north from the mouth of the Great Miami, was organized into the Territory of Indiana; which was divided, and from which Illinois Territory was formed in 1809. In June, 1816, a constitution was adopted, and at the ensuing session of Congress, Indiana was made a State.

General Remarks. The importance of Indiana, as a desirable State for the attention of the emigrant to the West, has been too much overlooked. Although not possessing quite equal advantages with Illinois, especially in the quality and amount of prairie soil, it is far superior to Ohio; and fully equal,—nay, in our estimation,—rather superior to Michigan. Almost every part is easy of access, and in a very few years the liberal system of internal improvements, adopted and in progress, will make almost every county accessible to public conveyances, and furnish abundant facilities to market.

Along the wide, alluvion bottoms of the streams, and amidst a rank growth of vegetation, there is usually more or less autumnal fever; yet, in general, there is very little difference in any of the Western States as to prospects of health.

Mechanics, school teachers, and laborers of every description, are much wanted in this State, as they are in all the States further west; and all may provide abundantly and easily all the necessaries of living for a family, if they will use industry, economy and sobriety.

CHAPTER XII.

ILLINOIS.

Boundaries and Extent—Face of the Country and Qualities of Soil—Inundated Land—River Bottoms, or Alluvion—Prairies—Barrens—Forest, or timbered Land—Knobs, Bluffs, Ravines and Sink-holes—Rivers, &c.—Animal, Mineral and Vegetable Productions—Manufactures—Civil Divisions—Tabular View of the Counties—Sketch of each County—Towns—Projected Improvements—Education—Government—General Remarks.

THE State of Illinois is situated between 37° and 42° 30′ north latitude; and between 10° 25′ and 14° 30′ west longitude from Washington city. It is bounded on the north by Wisconsin Territory, north-east by lake Michigan, east by Indiana, south-east and south by Kentucky, and west by Missouri. Its extreme length is three hundred and eighty miles; and its extreme width, two hundred and twenty miles; its average width, one hundred and fifty miles. The area of the whole State, including a small portion of lake Michigan within its boundaries, is 59,300 square miles.

The water area of the State is about 3750 square miles. With this, deduct 5550 square miles for irreclaimable wastes, and there remains 50,000 square miles, or 32,000,000 of acres of arable land in Illinois,—a much greater quantity than is found in any other State. In this estimate, inundated lands, submerged by high waters, but which may be reclaimed at a moderate expense, is included.

Face of the Country, and qualities of Soil. The general surface is level, or moderately undulating; the northern and southern portions are broken and somewhat hilly, but no portion of the State is traversed with ranges of hills or mountains. At the verge of the alluvial soil on the margins of rivers, there are ranges of "bluffs," intersected with ravines. The bluffs are usually from fifty to one hundred and fifty feet high, where an extended surface of table land commences, covered with prairies and forests of various shapes and sizes.

When examined minutely, there are several varieties in the surface of this State, which will be briefly specified and described.

1. *Inundated Lands.* I apply this term to allt hose portions, which, for some part of the year, are under water. These include portions of the river bottoms, and portions of the interior of large prairies, with the lakes and ponds which, for half the year or more, are without water. The term "bottom" is used throughout the West, to denote the alluvial

soil on the margin of rivers, usually called "intervales," in New England. Portions of this description of land are overflowed for a longer or shorter period, when the rivers are full. Probably one eighth of the bottom lands are of this description; for, though the water may not stand for any length of time, it wholly prevents settlement and cultivation, though it does not interrupt the growth of timber and vegetation. These tracts are on the bottoms of the Wabash, Ohio, Mississippi, Illinois and all the interior rivers.

When the rivers rise above their ordinary height, the waters of the smaller streams, which are backed up by the freshets of the former, break over their banks, and cover all the low grounds. Here they stand for a few days, or for many weeks, especially towards the bluffs; for it is a striking fact in the geology of the western country, that all the river bottoms are higher on the margins of the streams than at some distance back. Whenever increase of population shall create a demand for this species of soil, the most of it can be reclaimed at comparatively small expense. Its fertility will be inexhaustible, and if the waters from the rivers could be shut out by dykes or levees, the soil would be perfectly dry. Most of the small lakes on the American bottom disappear in the summer, and leave a deposit of vegetable matter undergoing decomposition, or a luxuriant coat of weeds and grass.

As our prairies mostly lie between the streams that drain the country, the interior of the large ones are usually level. Here are formed ponds and lakes, after the winter and spring rains, which remain, to be drawn off by evaporation, or absorbed by an adhesive soil. Hence the middle of our large, level prairies are wet, and for several weeks portions of them are covered with water. To remedy this inconvenience completely, and render all this portion of soil dry and productive, only requires a ditch or drain of two or three feet deep to be cut into the nearest ravine. In many instances, a single furrow with the plough would drain many acres. At present, this species of inundated land offers no inconvenience to the people, except in the production of miasm, and even that, perhaps, becomes too much diluted with the atmosphere to produce mischief before it reaches the settlements on the borders of the prairie. Hence the inference is correct, that our inundated lands present fewer obstacles to the settlement and growth of the country, and can be reclaimed at much less expense, than the swamps and salt marshes of the Atlantic States.

2. *River bottoms, or Alluvion.* The surface of our alluvial bottoms is not entirely level. In some places it resembles alternate waves of the ocean, and looks as though the waters had left their deposit in ridges, and retired.

The portion of bottom land capable of present cultivation, and on which the waters never stand, if, at an extreme freshet it is covered, is a soil of exhaustless fertility; a soil that for ages past has been gradually deposited by the annual floods. Its average depth on the American bottom, is from twenty to twenty-five feet. Logs of wood, and other indications, are found at that depth. The soil dug from wells on these bottoms, produces luxuriantly the first year.

The most extensive and fertile tract of this description of soil in this State, is the *American bottom*, a name it received when it constituted the western boundary of the United States, and which it has retained ever since. It commences at the mouth of the Kaskaskia river, five miles below the town of Kaskaskia, and extends northwardly, along the Mississippi, to the bluffs at Alton, a distance of ninety miles. Its average width is five miles, and contains about four hundred and fifty square miles, or 288,000 acres. Opposite St. Louis, in St. Clair county, the bluffs are seven miles from the river, and filled with inexhaustible beds of coal. The soil of this bottom is an argillaceous or a silicious loam, according as clay or sand happens to predominate in its formation.

On the margin of the river, and of some of its lakes, is a strip of heavy timber, with a thick undergrowth, which extends from half a mile to two miles in width; but from thence to the bluffs, it is principally prairie. It is interspersed with sloughs, lakes and ponds, the most of which become dry in autumn.

The soil of the American bottom is inexhaustibly rich. About the French towns it has been cultivated, and produced corn in succession for more than a century. without exhausting its fertilizing powers. The only objection that can be offered to this tract is its unhealthy character. This, however, has diminished considerably within eight or ten years. The geological feature noticed in the last article,—that all our bottoms are higher on the margin of the stream, than towards the bluffs,—explains the cause why so much standing water is on the bottom land, which, during the summer, stagnates and throws off noxious effluvia. These lakes are usually full of vegetable matter undergoing decomposition, and which produces large quantities of miasm. Some of the lakes are clear, with a sandy bottom, but the most are of a different character. The French settled near a lake or a river, apparently in the most unhealthy places, and yet their constitutions were little affected; and they usually enjoyed good health, though dwarfish and shriveled in their form and features.

"The villages of Kaskaskia, Prairie du Rocher and Cahokia, were built up by their industry, in places where Americans would have perished. Cultivation has, no doubt, rendered this tract more salubrious than formerly; and an increase of it, together with the construction of drains and canals, will make it one of the most eligible in the States. The old inhabitants advise the emigrants not to plant corn in the immediate vicinity of their dwellings, as its rich and massive foliage prevents the sun from dispelling the deleterious vapors."*

These lakes and ponds could be drained at a small expense, and the soil would be susceptible of cultivation. The early settlements of the Americans were either on this bottom, or the contiguous bluffs.

Besides the American bottom, there are others that resemble it in its general character, but not in extent. In Union county, there is an extensive bottom on the borders of the Mississippi. Above the mouth of the Illinois, and along the borders of the counties of Calhoun, Pike and Adams, there are a series of bottoms, with much good and elevated land; but the inundated grounds around present objections to a dense population at present.

The bottoms of Illinois, where not inundated, are equal in fertility, and the soil is less adhesive than most parts of the American bottom. This is likewise the character of the bottoms in the northern parts of the State.

The bottoms of the Kaskaskia are generally covered with a heavy growth of timber, and in many places inundated, when the river is at its highest floods.

* Beck

The extensive prairies adjoining, will create a demand for all this timber. The bottom lands on the Wabash are of various qualities. Near the mouth, much of it is inundated; higher up, it overflows in high freshets.

These bottoms, especially the American, are the best regions in the United States for raising stock, particularly horses, cattle and swine. Seventy-five bushels of corn to the acre is an ordinary crop. The roots and worms of the soil, the acorns and other fruits from the trees, and the fish of the lakes, accelerate the growth of swine. Horses and cattle find exhaustless supplies of grass in the prairies; and pea-vines, buffalo-grass, wild oats, and other herbage in the timber, for summer range; and often throughout most of the winter. In all the rush bottoms, they fatten during the severe weather, on rushes. The bottom soil is not so well adapted to the production of small grain, as of maize or Indian corn, on account of its rank growth, and being more subject to blast or fall down before harvest, than on the uplands.

3. *Prairies.* Much the largest proportion is undulating, dry and extremely fertile. Other portions are level; and the soil in some cases proves to be wet;—the water, not running off freely, is left to be absorbed by the soil, or evaporated by the sun. Craw-fish throw up their hillocks in this soil, and the farmer who cultivates it, will find his labors impeded by the water.

In the southern part, that is, south of the national road leading from Terre Haute to the Mississippi, the prairies are comparatively small, varying in size from those of several miles in width, to those which contain only a few acres. As we go northward, they widen and extend on the more elevated ground between the water-courses, to a vast distance, and are frequently from six to twelve miles in width. Their borders are by no means uniform. Long points of timber project into the prairies, and line the banks of the streams, and points of prairie project into the timber between these streams. In many instances are copses and groves of timber, from one hundred to two thousand acres, in the midst of prairies, like islands in the ocean. This is a common feature in the country between the Sangamon river and lake Michigan, and in the northern parts of the State. The lead-mine region, both in this State and Wisconsin Territory, abound with these groves.

The *origin* of these prairies has caused much speculation. We might as well dispute about the origin of forests, upon the assumption that the natural covering of the earth was grass. Probably one half of the earth's surface, in a state of nature, was prairies or barrens. Much of it, like our western prairies, was covered with a luxuriant coat of grass and herbage. The *steppes* of Tartary, the *pampas* of South America, the *savannas* of the Southern, and the *prairies* of the Western States, designate similar tracts of country. Mesopotamia, Syria and Judea had their ancient prairies, on which the patriarchs fed their flocks. Missionaries in Burmah, and travelers in the interior of Africa, mention the same description of country. Where the tough sward of the prairie is once formed, timber will not take root. Destroy this by the plough, or by any other method, and it is soon converted into forest land. There are large tracts of country in the older settlements, where thirty or forty years since, the farmers mowed their hay, that are now covered with a forest of young timber of rapid growth.

The fire annually sweeps over the prairies, destroying the grass and herbage, blackening the surface, and leaving a deposit of ashes to enrich the soil.

4. *Barrens.* This term, in the western dialect, does not indicate *poor land*, but a species of surface of a mixed character, uniting forest and prairie.

The timber is generally scattering, of a rough and stunted appearance, interspersed with patches of hazle and brushwood, and where the contest between the fire and timber is kept up, each striving for the mastery.

In the early settlements of Kentucky, much of the country below and south of Green river, presented a dwarfish and stunted growth of timber, scattered over the surface, or collected in clumps, with hazle and shrubbery intermixed. This appearance led the first explorers to the inference that the soil itself must necessarily be poor, to produce so scanty a growth of timber, and they gave the name of *barrens* to the whole tract of country. Long since, it has been ascertained that this description of land is amongst the most productive soil in the State. The term *barren* has since received a very extensive application throughout the West. Like all other tracts of country, the barrens present a considerable diversity of soil. In general, however, the surface is more uneven or rolling than the prairies, and sooner degenerates into ravines and sink-holes. Wherever timber barely sufficient for present purposes, can be found, a person need not hesitate to settle in the barrens. These tracts are almost invariably healthy; they possess a greater abundance of pure springs of water, and the soil is better adapted for all kinds of produce, and all descriptions of seasons, wet and dry, than the deeper and richer mould of the bottoms and prairies.

When the fires are stopped, these barrens produce timber, at a rate of which no northern emigrant can have any just conception. Dwarfish shrubs, and small trees of oak and hickory are scattered over the surface, where for years they have contended with the fires for a precarious existence, while a mass of roots, sufficient for the support of large trees, have accumulated in the earth. As soon as they are protected from the ravages of the annual fires, the more thrifty sprouts shoot forth, and in ten years are large enough for corn-cribs and stables.

As the fires on the prairies become stopped by the surrounding settlements, and the wild grass is eaten out and trodden down by the stock, they begin to assume the character of barrens; first, hazle and other shrubs, and finally, a thicket of young timber, covers the surface.

5. *Forest, or timbered Land.* In general, Illinois is abundantly supplied with timber, and were it equally distributed through the State, there would be no part in want. The apparent scarcity of timber where the prairie predominates, is not so great an obstacle to the settlement of the country as has been supposed. For many of the purposes to which timber is applied, substitutes are found. The rapidity with which the young growth pushes itself forward, without a single effort on the part of man to accelerate it, and the readiness with which the prairie becomes converted into thickets, and then into a forest of young timber, shows, that, in another generation, timber will not be wanting in any part of Illinois.

The kinds of timber most abundant are oak of various species, black and white walnut, ash of several kinds, elm, sugar-maple, honey-locust, hackberry, linden, hickory, cotton-wood, pecaun, mulberry, buckeye, sycamore, wild cherry, box-elder, sassafras, and persimmon. In the southern and eastern parts of the State are yellow poplar and beech; near the Ohio are cypress, and in several counties are clumps of yellow pine and cedar. On the Calamick, near the south end of lake Michigan, is a small forest of white pine. The undergrowth are red-bud, pawpaw, sumach, plum, crab-apple, grape vines, dogwood, spice-bush, green brier, hazle, &c. The alluvial soil of the rivers produces cotton-wood and sycamore timber of amazing size.

For ordinary purposes there is now timber enough in most parts of the State, to say nothing about the artificial production of timber, which may be effected with little trouble and expense. The black locust, a native of Ohio and Kentucky, may be raised from the seed, with less labor than a nursery of apple trees. It is of rapid growth, and, as a valuable and lasting timber, claims the attention of our farmers. It forms one of the cleanliest and most beautiful shades, and when in blossom, gives a rich prospect, and sends abroad a delicious fragrance.

6. *Knobs, Bluffs, Ravines and Sink-holes.* Under these heads are included tracts of uneven country found in various parts of the State.

Knobs are ridges of flint limestone, intermingled and covered with earth, and elevated one or two hundred feet above the common surface. This species of land is of little value for cultivation, and usually has a sprinkling of dwarfish, stunted timber, like the barrens.

The steep hills and natural mounds that border the alluvions have obtained the name of *bluffs*. Some are in long, parallel ridges, others are in the form of cones and pyramids. In some places, precipices of limestone rock, from fifty to one or two hundred feet high, form these bluffs.

Ravines are formed amongst the bluffs, and often near the borders of prairies, which lead down to the streams.

Sink-holes are circular depressions in the surface, like a basin. They are of various sizes, from ten to fifty feet deep, and from ten to one or two hundred yards in circumference. Frequently they contain an outlet for the water received by the rains. Their existence shows that the substratum is secondary limestone, abounding with subterraneous cavities.

There are but few tracts of stony ground in the State; that is, where loose stones are scattered over the surface, and imbedded in the soil. Towards the northern part of the State, tracts of stony ground exist. Quarries of stone exist in the bluffs and in the banks of the streams and ravines throughout the State. The soil is porous, easy to cultivate, and exceedingly productive. A strong team is required to break up the prairies, on account of the firm, grassy sward which covers them; but when subdued, they become fine, arable lands.

Rivers, &c. This State is surrounded and

intersected by navigable streams. The Mississippi, Ohio and Wabash rivers are on three sides; the Illinois, Kaskaskia, Sangamon, Muddy, and many smaller streams are entirely within its borders; and the Kankakee, Fox, Rock, and Vermilion of the Wabash run part of their course within this State. The Mississippi meanders its western border for seven hundred miles. Its principal tributaries within Illinois, are Rock, Illinois, Kaskaskia and Muddy rivers. The Illinois river commences at the junction of the Kankakee, which originates near the South Bend, in Indiana, and the Des Plaines, which rises in Wisconsin Territory. From their junction, the Illinois runs nearly a west course (receiving Fox river at Ottawa, and Vermilion near the foot of the rapids), to Hennepin, where it curves to the south and then to the south-west, receiving a number of tributaries, the largest of which are Spoon river from the right and Sangamon from the left, till it reaches Naples. Here it bends gradually to the south, and continues that course till within six miles of the Mississippi, when it curves to the south-east, and finally, to nearly an eastern course. Its length (without reckoning the windings of the channel in navigation), is about two hundred and sixty miles, and is navigable for steam-boats, at a moderate stage of water, to the foot of the rapids. The large streams on the eastern side of the State are Iroquois, a tributary to the Kankakee, Vermilion of the

Wabash, which enters that river in Indiana, Embarras, that has its source near that of the Kaskaskia, runs south-easterly, and enters the Wabash, nine miles below Vincennes, and Little Wabash, near its mouth. Along the Ohio, the only streams deserving note are the Saline and Bay creeks, and Cash river, the last of which enters the Ohio six miles above its confluence with the Mississippi.

Productions. These are naturally classed into *mineral*, *animal* and *vegetable.*

Minerals. The northern portion of Illinois is inexhaustibly rich in mineral productions, while coal, secondary limestone, and sandstone, are found in every part.

Iron ore has been found in the southern parts of the State, and is said to exist in considerable quantities in the northern parts.

Native *copper*, in small quantities, has been found on Muddy river, in Jackson county, and back of Harrisonville, in the bluffs of Monroe county. Crystallized gypsum has been found in small quantities in St. Clair county. Quartz crystals exist in Gallatin county.

Silver is supposed to exist in St. Clair county, two miles from Rock Spring, from whence Silver creek derives its name. In early times a shaft was sunk here, by the French, and tradition tells of large quantities of the precious metals being obtained.

In the southern part of the State, several sections of land have been reserved from sale, on account of the silver ore they are supposed to contain.

Lead is found in vast quantities in the northern part of Illinois, and the adjacent Territory. Here are the richest lead mines hitherto discovered on the globe. This portion of country lies principally north of Rock river and south of the Wisconsin. Dubuque's and other rich mines, are west of the Mississippi.

Native copper, in large quantities, exists in this region, especially at the mouth of Plum creek, and on the Peek-a-ton-o-kee, a branch of Rock river.

The following is a list of the principal diggings in that portion of the lead-mine region that lies between Rock river and the Wisconsin, embracing portions of Illinois State, and Wisconsin Territory. Some of these diggings are, probably relinquished, and many new ones commenced:—

Apple creek,
GALENA and vicinity,
Cave diggings,
Buncombe,
Natchez,
Hardscrabble,
New diggings,
Gratiot's Grove,
Spulburg,
W. S. Hamilton's,
Cotttle's,
McNutt's,
Menomonee creek,
Plattsville,
CASSVILLE and vicinity,
Madden's,
Mineral point,
Dodgeville,
Worke's diggings,
Brisbo's,
Blue mounds,
Prairie springs,
Hammett & Campbell's,
Morrison's;

and many others.

Amount of Lead manufactured. For many years the Indians, and some of the French hunters and traders, had been accustomed to dig lead in these regions. They never penetrated much below the surface, but obtained considerable quantities of the ore which they sold to the traders.

In 1823, the late Colonel James Johnson, (brother to the Hon. R. M. Johnson,) of Great Crossings, Kentucky, obtained a lease from the United States government, and made arrangements to prosecute the business of smelting, with considerable force, which he did the following season. This attracted the attention of enterprising men in Illinois, Missouri, and other States. Some went on in 1826, more followed in 1827, and in 1828 the country was almost literally filled with miners, smelters, merchants, speculators, gamblers, and every description of character. Intelligence, enterprise and virtue were thrown in the midst of dissipation, gaming, and every species of vice. Such was the crowd of adventurers in 1829, to this hitherto almost unknown and desolate region, that the lead business was greatly overdone, and the market for a while nearly destroyed. Fortunes were made almost upon the turn of the spade, and lost with equal facility. The business has revived, and is profitable. Exhaustless qnantities of mineral exist here, over a tract of country two hundred miles in extent.

The following table shows the amount of

lead made annually at these diggings, from 1821, to September 30, 1835:—

Amount of Lead manufactured.

				Pounds.
From 1821, to Sept. 1823,..........				335,130
For the year ending Sept. 30,			1824,....	175,220
"	"	"	1825,....	664,530
"	"	"	1826,....	958,842
"	"	"	1827,....	5,182,180
"	"	"	1828,....	11,105,810
"	"	"	1829,....	13,344,150
"	"	"	1830,....	8,323,998
"	"	"	1831,....	6,381,900
"	"	"	1832,....	4,281,876
"	"	"	1833,....	7,941,792
"	"	"	1834,....	7,971,579
"	"	"	1835,....	3,754,290
		Total,		70,420,357

The rent, accruing to government, for the same period, is a fraction short of six millions of pounds. The government formerly received ten per cent. in lead, for rent: now it is six per cent.

A part of the mineral land in Wisconsin Territory has been surveyed and brought into market, which will add greatly to the stability and prosperity of the mining business.

Coal. Bituminous coal abounds in Illinois. It may be seen, frequently, in the ravines and gullies, and in the points of bluffs. Exhaustless beds of this article exist in the bluffs of St. Clair county, bordering on the American bottom, of which large quantities are transported to St. Louis, for fuel. There is scarcely a county in the State, but what can furnish coal, in reasonable quantities. Large beds are said to exist near the Vermilion of the Illinois, and in the vicinity of the rapids of the latter.

Agatized Wood. A petrified tree, of black walnut, was found in the bed of the river Des Plaines, about forty rods above its junction with the Kankakee, imbedded in a horizontal position, in a stratum of sandstone. There is fifty-one and a half feet of the trunk visible; eighteen inches in diameter at its smallest end, and probably three feet at the other end.

Muriate of Soda, or common salt. This is found in various parts of the State, held in solution in the springs. The mauufacture of salt, by boiling and evaporation, is carried on in Gallatin county, twelve miles west-northwest from Shawneetown; in Jackson county, near Brownsville; and in Vermilion county, near Danville. The springs and land are owned by the State, and the works leased.

A coarse freestone, much used in building, is dug from quarries near Alton, on the Mississippi, where large bodies exist.

Scattered over the surface of our prairies, are large masses of rock, of granatic formation, roundish in form, usually called by the people, *lost rocks*. They will weigh from one thousand to ten or twelve thousand pounds, are entirely detached, and frequently are found several miles distant from any quarry. Nor has there ever been a quarry of granite discovered in the State. These stones are denominated *bowlders*, in mineralogy; they usually lie on the surface, or are partially imbedded in the soil of our prairies, which is unquestionably of diluvial formation. How they came here is a question of difficult solution.

Medicinal waters are found in different parts of the State. These are chiefly sulphur springs and chalybeate waters. There is said to be one well in the southern part of the State, strongly impregnated with the sulphate of magnesia or Epsom salts, from which considerable quantities have been made for sale, by simply evaporating the water, in a kettle, over a common fire.

There are several sulphur springs in Jefferson county, to which persons resort for health.

Vegetable Productions. The principal trees and shrubs of Illinois have been noticed under the head of *forest or timbered land*. Of oak there are several species, as over-cup, burr-oak, swamp or water oak, white oak, red or Spanish oak, post oak and black oak of several varieties, with the black-jack, a dwarfish gnarled-looking tree, excellent for fuel, but good for nothing else.

The black walnut is much used for building materials and cabinet work, and sustains a fine polish.

In most parts of the State, grape-vines, indigenous to the country, are abundant, which yield grapes that might advantageously be made into excellent wine. Foreign vines are susceptible of easy cultivation. These are cultivated to a considerable extent, at Vevay, Switzerland county, Indiana, and at New Harmony, on the Wabash. The indigenous vines are prolific, and produce excellent fruit. They are found in every variety of soil; interwoven in every thicket in the prairies and barrens; and climbing to the tops of the very highest trees on the bottoms. The French, in early times, made so much wine as to export some to France; upon which the proper authorities prohibited the introduction of wine from Illinois, lest it might injure the sale of that staple article of the kingdom. I think the act was passed by the Board of Trade, in 1774. The editor of the Illinois Magazine remarks, "We know one gentleman who made twenty-seven barrels of wine in a single season, from the grapes gathered with but little labor, in his immediate neighborhood."

The wild plum is found in every part of the State; but in most instances the fruit is too sour for use, unless for preserves. Crab-apples are equally prolific, and make fine preserves, with about double their bulk of sugar. Wild cherries are equally productive. The persimmon is a delicious fruit, after the frost has destroyed its astringent properties. The black mulberry grows in most parts, and is used for the feeding of silk-worms, with success. They appear to thrive and spin as well as on the Italian mulberry. The gooseberry, strawberry and blackberry, grow wild and in great profusion. Of nuts, the hickory, black walnut and pecaun, deserve notice. The last is an oblong, thin-shelled, delicious nut, that grows on a large tree, a species of the hickory (the *Carya olivæformis* of Nuttall). The pawpaw grows on the bottoms and rich, timbered uplands, and produces a large, pulpy and luscious fruit. Of domestic fruits, the apple and peach are chiefly cultivated. Pears are tolerably plenty in the French settlements, and quinces are cultivated with success by some Americans. Apples are easily cultivated, and are very productive. The trees can be made to bear fruit to considerable advantage, in seven years, from the seed. Many varieties are of fine flavor, and grow to a large size. I have measured apples, the growth of St. Clair county, that exceeded thirteen inches in circumference. Some of the early American settlers provided orchards; they now reap the advantages. But a large proportion of the population of the frontiers are content without this indispensable article in the comforts of a yankee farmer. Cider is made in small quantities in the old settlements. In a few years, a supply of this beverage can be had in most parts of Illinois.

Peach trees grow with great rapidity, and decay proportionably soon. From ten to fifteen years may be considered the life of this tree. Our peaches are delicious, but they sometimes fail, by being destroyed in the germ, by winter frosts.—The bud swells prematurely.

Garden vegetables can be produced here in vast profusion, and of excellent quality.

That we have few of the elegant and well dressed gardens of gentlemen in the old States, is admitted; which is not owing to climate, or soil, but to the want of leisure and means.

Our Irish potatoes, pumpkins and squashes are inferior, but not our cabbages, peas, beets or onions.

A cabbage-head, two or three feet in diameter, including the leaves, is no wonder on this soil. Beets often exceed twelve inches in circumference. Parsnips will penetrate our light, porous soil, to the depth of two or three feet.

The *cultivated vegetable productions in the field*, are maize or Indian corn, wheat, oats, barley, buckwheat, Irish potatoes, sweet potatoes, turnips, rye for horse-feed and distilleries, tobacco, cotton, hemp, flax, the castor bean, and every other production common to the Middle States.

Maize is a staple production. No farmer can live without it, and hundreds raise little else. This is chiefly owing to the ease with which it is cultivated. Its average produce is fifty bushels to the acre. I have oftentimes seen it produce seventy-five bushels to the acre, and in a few instances, exceed one hundred.

Wheat yields a good and sure crop, especially in the counties bordering on the Illinois river. It weighs upwards of sixty pounds per bushel; and flour from this region has preference in the New Orleans market, and passes better inspection than the same article from Ohio or Kentucky.

In 1825, the weevil, for the first time, made its appearance in St. Clair and the adjacent counties, and has occasionally renewed its visits since. Latterly, some fields have been injured by the fly.

A common but slovenly practice amongst our farmers, is to sow wheat amongst the standing corn in September, and cover it, by running a few furrows with the plough between the rows of corn. The dry stalks are then cut down in the spring, and left on the ground. Even by this imperfect mode, fifteen or twenty bushels of wheat to the acre, are produced. But where the ground is duly prepared by fallowing, and the seed put in at the proper time, a good crop, averaging from twenty-five to thirty-five bushels per acre rarely fails to be procured.

The average price of wheat, at present, is a dollar per bushel, varying a little according to the competition of mills and facilities to market. In many instances a single crop of wheat will more than pay the expenses of purchasing the land, fencing, breaking the prairie, seed, putting in the crop, harvesting, threshing and taking it to market. Wheat is now frequently sown on the prairie land, as a first crop, and a good yield obtained.

Flouring-mills are now in operation in many of the wheat-growing counties. Steam power is getting into extensive use, both for sawing timber and manufacturing flour.

It is to be regretted, that so few of our farmers have erected barns for the security of their crops. No article is more profitable, and really more indispensable to a farmer, than a large barn.

Oats have not been much raised till lately. They are very productive, often yielding from forty to fifty bushels on the acre, and usually sell for twenty-five cents per bushel. The demand, for the use of stage and travelers' horses, is increasing.

Hemp is an indigenous plant, in the southern part of this State, as it is in Missouri. It has not been extensively cultivated; but wherever tried, is found very productive, and of an excellent quality. It might be made a staple of the country.

Tobacco, though a filthy and noxious weed, which no human being ought ever to use, can be produced in any quantity, and of the first quality, in Illinois.

Cotton, for many years, has been successfully cultivated in this State, for domestic use, and some for exportation. Two or three spinning-factories are in operation, and produce cotton yarn, from the growth of the country, with promising success. This branch of business admits of enlargement, and invites the attention of eastern manufacturers with small capital. Much of the cloth made in families who have emigrated from States south of the Ohio, is from the cotton of the country.

Flax is produced and of a tolerable quality, but not equal to that of the Northern States. It is said to be productive and good in the northern counties.

Barley yields well, and is a sure crop.

The *palma christi*, or castor-oil bean, is produced in considerable quantities, in Madison, Randolph and other counties, and large quantities of oil are expressed and sent abroad.

Sweet potatoes are a delicious root, which yields abundantly, especially on the American bottom and rich, sandy prairies.

But little has been done to introduce cultivated *grasses*. The prairie grass looks coarse and unsavory, and yet our horses and cattle will thrive well on it.

To produce timothy with success, the ground must be well cultivated in the summer, either by an early crop, or by fallowing, and the seed sown about the twentieth of September, at the rate of *ten or twelve quarts of clean seed to the acre*, and lightly brushed in. If the season is in any way favorable, it will get a rapid start before winter. By the last week in June, it will produce two tons per acre, of the finest hay. It then requires a dressing of stable or yard-manure, and occasionally the turf may be scratched with a harrow, to prevent the roots from binding too hard. By this process, timothy meadows may be made and preserved. There are meadows in St. Clair county, which have yielded heavy crops of hay in succession, for several years, and bid fair to continue, for an indefinite period. Cattle, and especially horses, should never be permitted to run in meadows, in Illinois. The fall grass may be cropped down by calves and colts. There is but little more labor required to produce a crop of timothy, than a crop of oats; and as there is not a stone or a pebble to interrupt, the soil may be turned up every third or fourth year, for corn, and afterwards laid down to grass again.

A species of blue grass is cultivated by some farmers, for pastures. If well set, and not eaten down in summer, blue grass pastures may be kept green and fresh till late in autumn, or even in the winter. The English spire-grass has been cultivated with success in the Wabash county.

Of the trefoil, or clover, there is but little cultivated. A prejudice exists against it, as it is imagined to injure horses, by affecting the glands of the mouth, and causing them to slaver. It grows luxuriantly, and may be cut for hay, early in June. The white clover comes in naturally, where the ground has been cultivated and thrown by, or along the sides of old roads and paths. Clover pastures would be excellent for swine.

Animals. Of *wild animals* there are several species. The *buffalo* is not found on this side of the Mississippi, nor within several hundred miles of St. Louis. This animal once roamed at large over the prairies of Illinois, and was found in plenty, thirty-five years since.

Wolves, *panthers* and *wild-cats* still exist on the frontiers and through the unsettled portions of the country, and annoy the farmer, by destroying his sheep and pigs.

Deer are also very numerous, and are valuable, particularly to that class of our population which has been raised to frontier habits; the flesh affording them food, and the skins clothing. Fresh venison-hams usually sell for twenty-five cents each, and when properly cured, are a delicious article. Many of the frontier people dress the skins and make them into pantaloons and hunting-shirts. These articles are indispensable to all who have occasion to travel, in viewing land, or for any other purpose beyond the settlements, as cloth garments, in the shrubs and vines, would soon be in strings.

It is a novel and pleasing sight to a stranger, to see the deer, in flocks of eight, ten, or fifteen in number, feeding on the grass of the prairies, or bounding away, at the sight of a traveler.

The *brown bear* is also an inhabitant of the unsettled parts of this State, although he is continually retreating before the advance of civilization.

Foxes, *racoons*, *opossums*, *gophars* and *squirrels* are also numerous, as are *muskrats*, *otters*, and occasionally *beaver*, about our rivers and lakes. Racoons are very common, and frequently do mischief, in the fall, to the corn. Opossums sometimes trouble the poultry.

The *gophar* is a singular little animal, about the size of a squirrel, which burrows in the ground, and is seldom seen; but its *works* make it known. It labors during the night, in digging subterranean passages in the rich soil of the prairies, and throws up hillocks of fresh earth, within a few feet distance from each other, and from twelve to eighteen inches in height.

The gray and fox *squirrels* often do mischief in the cornfields, and the hunting of them makes fine sport for the boys.

Common rabbits exist in every thicket, and annoy nurseries and young orchards exceedingly. The fence around a nursery must always be so close as to shut out rabbits; and young apple trees must be secured, at the approach of winter, by tying straw or corn-stalks around their bodies, for two or three feet in height, or the bark will be stripped off by these mischievous animals.

Wild horses are found ranging the prairies and forests in some parts of the State. They are small in size, of the Indian or Canadian breed and very hardy. They are found chiefly in the lower end of the American bottom, near the junction of the Kaskaskia and Mississippi rivers, called *the point.* They are the offspring of the horses brought there by the first settlers, and which were suffered to run at large. The Indians of the West have many such horses, which are commonly called Indian ponies.

Domestic Animals. These are the same as are found in other portions of the United States. But little has been done to improve the breed of horses among us. Our common riding or working-horses average about fifteen hands in height. They are much more used here than in the Eastern States, and many a farmer keeps half a dozen or more. Much of the traveling throughout the western country, both by men and women, is performed on horseback; and a large proportion of the land carriage is by means of large wagons, with from four to six stout horses for a team. A great proportion of the ploughing is performed by horse-labor. Horses are more subject to diseases in this country than in the old States, which is thought to be occasioned by bad management, rather than by the climate. A good farm-horse can be purchased for fifty dollars. Riding or carriage-horses, of a superior quality, cost from seventy-five to eighty dollars. Breeding-mares are profitable stock for every farmer to keep, as their annual expense in keeping is but trifling: their labor is always needed, and their colts, when grown, find a ready market. Some farmers keep a stallion and eight or ten brood-mares.

Mules are brought into Missouri, and find their way to Illinois, from the Mexican dominions. They are a hardy animal, grow to a good size, and are used by some, both for labor and riding.

Our *neat cattle* are usually inferior in size to those of the old States. This is owing entirely to bad management. Our cows are not penned up in pasture fields, but suffered to run at large over the commons. Hence *all* the calves are preserved, without respect to quality, to entice the cows homeward at evening. In autumn their food is very scanty, and during the winter they are permitted to pick up a precarious subsistence amongst fifty or a hundred head of cattle. With such management, is it surprising that our cows and steers are much inferior to those of the old States? And yet, our beef is the finest in the world. It bears the best inspection of any in the New Orleans market. By the first of June, and often by the middle of May, our young cattle on the prairies are fit for market. They do not yield large quantities of tallow, but the fat is well proportioned throughout the carcass, and the meat tender and delicious. By inferiority, then, I mean the *size* of our cattle in general, and the quantity and quality of the milk of cows.

Common cows, if suffered to lose their milk in August, become sufficiently fat for table use by October. Farrow-heifers and steers are good beef, and fit for the knife at any period after the middle of May. Nothing is more common than for an Illinois farmer to go among his stock, select, shoot down, and dress a fine beef, whenever fresh meat is needed. This is often divided out amongst the neighbors, who, in turn, kill and share likewise. It is common at camp and other large meetings, to kill a beef and three or four hogs for the subsistence of friends from a distance..

Steers from three years old or more, have been purchased in great numbers in Illinois, by drovers from Ohio. Cattle are sometimes sent in flat-boats down the Mississippi and Ohio, for the New Orleans market.

We can hardly place limits upon the amount of beef cattle that Illinois is capable of producing. A farmer calls himself poor, with a hundred head of horned cattle around him. A cow in the spring is worth from seven to ten or fifteen dollars. Some of the best quality will sell higher. And let it be distinctly understood, once for all, that a poor man can always purchase horses, cattle, hogs, and provisions, for labor, either by the day, month, or job.

Cows, in general, do not produce the same amount of milk, nor of as rich a quality as in older States. Something is to be attributed to the nature of our pastures, and the warmth of our climate, but more to causes already assigned. If ever a land was characterized justly, as "flowing with milk and honey," it is Illinois and the adjacent States. From the springing of the grass till September, butter is made in great profusion. It sells at that season, in market, for about ten cents. With proper care, it can be preserved in tolerable sweetness for winter use. Late in autumn and early in the winter, sometimes butter is not plenty. The feed becomes dry, the cows range further off, and do not come up readily for milking, and dry up. A very little trouble would enable a farmer to keep three or four good cows in fresh milk at the season most needed.

Cheese is made by many families, especially in the counties bordering on the Illinois river. Good cheese sells for eight and sometimes ten cents, and finds a ready market.

Swine. This species of stock may be called a staple in the provision of Illinois. Thousands of hogs are raised without any expense, except a few breeders to start with, and a little attention in hunting them on the range, and keeping them tame.

Pork that is made in a domestic way, and fatted on corn, will sell from three to four and five dollars, according to size, quality, and the time when it is delivered. With a pasture of clover or blue grass, a well-filled corn crib, a dairy, and slop barrel, and the usual care that a New Englander bestows on his pigs, pork may be raised from the sow, fatted, and killed, and weigh from two hundred to two hundred and fifty, within twelve months; and this method of raising pork would be profitable.

Few families in the West and South put up their pork in salt pickle. Their method is to salt it sufficiently to prepare it for smoking, and then make bacon of hams, shoulders, and middlings or broadsides. The price of bacon, taking the hog round, is about seven and eight cents. Good hams command eight and ten cents in the St. Louis market. Stock hogs, weighing from sixty to one hundred pounds, alive, usually sell from one to two dollars per head. Families consume much more meat in the West, in proportion to numbers, than in the old States.

Sheep do very well in this country, especially in the older settlements, where the grass has become short, and they are less molested by wolves.

Poultry is raised in great profusion; and large numbers of fowls taken to market.

Ducks, geese, swans, and many other aquatic birds, visit our waters in the spring. The small lakes and sloughs are often literally covered with them. Ducks, and some of the rest, frequently stay through the summer and breed.

The prairie fowl is seen in great numbers on the prairies in the summer, and about the cornfields, in the winter. This is the grouse of the New York market. They are easily taken in the winter.

Partridges (the quail of New England), are taken with nets, in the winter, by hundreds in a day, and furnish no trifling item in the luxuries of the city market.

Bees. These laborious and useful insects, are found in the trees of every forest.

Many of the frontier people make it a prominent business, after the frost has killed the vegetation, to hunt them for the honey and wax, both of which find a ready market. Bees are profitable stock for the farmer, and are kept to a considerable extent.

Silk-worms are raised by a few persons. They are capable of being produced to any extent, and fed on the common black mulberry of the country.

Salt. The principal salines of this State have been mentioned under the head of minerals. The principal works are at Gallatin, Big Muddy, and Vermilion salines.

Steam Mills for flouring and sawing are becoming very common, and in general are profitable. Some are now in operation with four run of stones, and which manufacture one hundred barrels of flour in a day. Mills propelled by steam, water, and animal power, are constantly increasing. Steam mills will become numerous, particularly in the southern and middle portions of the State; and it is deserving remark, that, while these portions are not well supplied with durable water power, they contain, in the timber of the forest, and the inexhaustible bodies of bituminous coal, abundant supplies of fuel; while the northern portion, though deficient in fuel, has abundant water power. A good steam saw-mill with two saws, can be built for $1500; and a steam flouring-mill with two run of stones, elevators and other apparatus complete, and of sufficient force to turn out forty or fifty barrels of flour per day, may be built for from $3500 to $5000. Ox-mills, on an inclined plane, and horse-mills, by draught, are common through the country.

Castor oil. Considerable quantities of this article have been manufactured in Illinois, from the *palma christi*, or castor-bean. One bushel of the beans will make nearly two gallons of the oil. There are five or six castor-oil presses in the State; in Madison, Randolph, Edwards, and perhaps in other counties. Mr. Adams, of Edwardsville, in 1825, made five hundred gallons, which then sold at the rate of $2 50 per gallon;—in 1826, he made eight hundred gallons;—in 1827, one thousand gallons, the price then, $1 75;—in 1828, one thousand eight hundred gallons, price $1:—in 1830, he started two presses, and made upwards of ten thousand gallons, which sold for from seventy-five to eighty-seven cents per gallon;—in 1831, about the same quantity. That and the following season being unfavorable for the production of the bean, there has been a falling off in the quantity. The amount manufactured in other parts of the State has probably exceeded that made by Mr Adams.

Lead. In Jo Daviess county are eight or ten furnaces for smelting lead. The amount of this article made annually at the mines of the Upper Mississippi, has been given under the head of minerals.

Manufactures. In the infancy of a State, little can be expected in machinery and manufactures. And in a region so much deficient in water power as some parts of Illinois are, still less may be looked for; yet Illinois is not entirely deficient in manufacturing enterprise.

There is in this State, as in all the Western States, a large amount of domestic manufactures made by families. All the trades, nedful to a new country, are in existence. Carpenters, wagon-makers, cabinet-makers, blacksmiths, tanners, &c., may be found in every county and town, and thousands more are wanted.

There has been a considerable falling off in the manufacture of whisky within a few years, and it is sincerely hoped by thousands of our citizens that this branch of business, so decidedly injurious to the morals and happiness of communities and individuals, will entirely decline. Several companies, for manufacturing purposes, have been incorporated by the legislature.

Boat-building will soon become a branch of business in this State. Some steam-boats have been constructed already within this State, along the Mississippi. It is thought that Alton and Chicago are convenient sites for this business.

Civil Divisions. There are sixty-six counties laid off in this State, fifty-nine of which are organized for judicial purposes. The counties of Will, Whiteside, Kane, Ogle, McHenry and Winnebago were laid off at the session of the legislature, Jan. 1836. The county of Will was formed from portions of Cook, Lasalle and Iroquois, with the town of

Juliet, near the junction of the Kankakee and Des Plaines, for its seat of justice.

In this State there are no civil divisions into townships, as in Ohio, Indiana, &c. The township tracts of six miles square, in the public surveys, relate exclusively to the land system. The State is divided into three districts to elect representatives to Congress, and into six circuits for judicial purposes.

Tabular View of the Counties.

Counties.	Date of Formation.	Square miles.	Votes. 1834.	Population in 1835.	Seats of Justice.	Bearing and distance frm Vandalia.
Adams,	1825	8[illegible]0	7[illegible]8	7042	Quincy,	175 *n.w.*
Alexander,	1819	375	249	2050	Unity,	135 *s.*
Bond,	1817	360	519	3580	Greenville,	19 *w.s.w.*
Calhoun,	1825	260	151	1091	Gilead,	134 *w.n.w.*
Champaign,	1833	864	102	1045	Urbanna,	103 *n.n.e.*
Clark,	1819	500	451	3413	Darwin,* or Marshall,	82 *e.n.e.*
Clay,	1824	620	172	1648	Maysville,	50 *s e.*
Clinton	1824	500	414	2648	Carlyle,	28 *s.s.w.*
Crawford,	1816	378	519	3540	Palestine,	100 *e.*
Coles,	1830	1248	680	5125	Charleston,	75 *n.e.*
Cook,	1830	†	528	98[illegible]6	Chicago,	2[illegible]8 *n.n.e.*
Edgar,	1823	648	788	6668	Paris,	100 *n.e.*
Edwards,	1814	200	239	2006	Albion,	96 *s.e.*
Effingham,	1831	486	129	1055	Ewington,	29 *e.n.e.*
Fayette,	1821	684	665	3638	VANDALIA,	
Franklin,	1818	850	759	5551	Frankfort,	83 *s.*
Fulton,	1825	590	607	5917	Lewistown,	135 *n.n.w.*
Gallatin,	1812	828	1312	8660	Equality,	100 *s.s.e.*
Greene,	1821	912	1360	12[illegible]74	Carrollton,	90 *w.n.w.*
Hamilton,	1821	378	460	2877	McLeansboro',	76 *s.s e.*
Hancock,	1825	775	357	3249	Carthage,	180 *n.w.*
Henry, (not organized,	1825	800	—	118		210 *n.n.w.*
Iroquois,	1833	†	67	1164	(Not established),	165 *n.n.e.*
Jackson,	1816	576	354	2783	Brownsville,	96 *s.s.w.*

* It is expected the seat of justice of Clark county will be removed to *Marshall*, ten miles north-west from Darwin, and on the national road. The distance is computed to Marshall.

Counties.	Date of Formation.	Square miles.	Votes, 1834.	Population in 1830.	Seats of Justice.	Bearing and distance frm Vandalia.
Jasper,........	1831	288	—	415	Newton,.........	60 *e.*
Jefferson,.....	1819	576	455	3359	Mount Vernon,...	48 *s.s.e.*
Jo Daviess, ...	1827	†	492	4038	Galena,..........	300 *n n.w.*
Johnson,......	1812	486	316	2166	Vienna,..........	120 *s.*
Kane,‡........	1836	—	—	—		
Knox, ...	1825	792	180	1600	Knoxville, ...	182 *n.n.w.*
Lasalle, ...	1831	†	289	4754	Ottawa, ...	187 *n.*
Lawrence, ...	1821	560	618	4450	Lawrenceville, ...	88 *e.s.e.*
Macon,.......	1829	1404	292	3022	Decatur, ...	75 *n.*
Madison,......	1812	750	1307	9016	Edwardsville, ...	58 *w.*
Macoupen,....	1829	720	624	5554	Carlinville, ...	55 *w.n.w.*
Marion,.......	1823	576	372	2844	Salem, ...	95 *s.s.e.*
McDonough, ..	1825	576	304	2883	Macomb, ...	155 *n.w.*
McHenry,‡....	1836	—	—	—		
McLean, ...	1830	1916	496	5311	Bloomington, ...	120 *n.*
Mercer,.......	1825	558	—	497	New Boston,.....	209 *n.w.*
Monroe, ...	1816	360	449	2660	Waterloo, ...	72 *s.w.*
Montgomery, ..	1821	960	475	3740	Hillsboro',.......	28 *n.w.*
Morgan, ...	1823	1150	2717	19214	Jacksonville, ...	91 *n.w.*
Ogle,‡........	1836	—	—	—		
Peoria,........	1825	648	2[illegible]3	3220	Peoria, ...	141 *n.n.w.*
Perry, ...	1827	446	273	2201	Pinckneyville,....	71 *s.s.w.*
Pike,.........	1821	800	657	6037	Pittsfield,........	126 *w.n.w.*
Pope,.........	1816	576	444	3756	Golconda,........	1[illegible]0 *s.s.e.*
Putnam, ...	1825	1340	383	4021	Hennepin,........	180 *n.*
Randolph, ...	1795	540	814	5695	Kaskaskia,.......	99 *s.s.w.*
Rock Island, ..	1831	377	83	616	Stephenson,......	220 *n.w.*
Sangamon,....	1821	1234	2219	17573	Springfield, ...	79 *n.n.w.*
Schuyler, ...	1825	864	680	6361	Rushville,........	128 *n.w.*
Shelby, ...	1827	1080	636	4848	Shelbyville, ...	40 *n.n.e.*
St. Clair, ...	1795	1330	1183	9055	Bellville, ...	64 *w.s.w.*
Tazewell,.....	1827	1130	433	5850	Tremont, ...	131 *n.*
Union,........	1818	396	545	4156	Jonesboro',.......	120 *s.*
Vermilion.....	1826	1000	1025	8 03	Danville, ...	135 *n.e.*
Wabash,......	1824	180	441	3010	Mount Carmel,...	95 *s.e.*
Warren, ...	1825	900	266	2623	Monmouth, ...	184 *n.w.*
Washington, ..	1818	656	333	3092	Nashville, ...	48 *s.s.w.*
Wayne,.......	1819	576	471	2939	Fairfield,........	76 *s e.*
White, ...	1815	516	977	6489	Carmi,...........	103 *s.e.*
Whiteside,‡...	1836	—	—	—		
Will,‡........	1836	—	—	—	Juliet, ...	
Winnebago,‡..	1836	—	—	—		

† These counties have been recently subdivided, and their superficial area is not known.

‡ These counties were formed January, 1836, and were taken from Jo Daviess, Lasalle, Cook and Iroquois. The seats of justice not established, and much of the land unsurveyed, though rapidly settling.

The following is a sketch of each county in the State:

Adams.—The streams are Bear creek and branches, Cedar, Tyrer, Mill, Fall, and Pigeon creeks, with the Mississippi river on its western border. Timber various, with equal portions of prairie. First rate county.

Alexander.—In the forks of the Ohio and Mississippi, with Cash river through it. All timbered,—half alluvion,—some inundated at high water,—lime and sandstone on the Ohio;—soil, generally rich.

Bond.—Shoal creek and its branches through it, with Hurricane creek on the east side;—proportioned into timber and prairie; rather level; second rate. Sandstone, coal, and salt springs.

Calhoun.—Long and narrow, in the forks of the Illinois and Mississippi;—alluvial and sometimes inundated along the rivers;—broken bluffs and interior table land;—good soil;—prairies at the foot of the bluffs. Coal, lime and sandstone.

Champaign.—The streams are the heads of the Kaskaskia, Sangamon, Vermilion of Illinois, Salt Fork of the Vermilion of the Wabash, and the Embarras, all running in opposite directions. Extensive prairies, a little undulating and rich;—timber in groves;—many granite bowlders.

Clark.—North Fork of Embarras, Mill and Big creeks. Timber and prairie,—second rate soil.

Clay.—Watered by Little Wabash and tributaries. Two thirds prairie,—of inferior quality,—rather level and wet.

Clinton.—Kaskaskia river, with its tributaries, Crooked, Shoal, Beaver and Sugar creeks pass through it. Equally proportioned into timber and prairie. Soil, second rate; surface, a little undulating.

Coles.—The Kaskaskia, Embarras, and heads of the Little Wabash water it. Much excellent land,—much undulating, rich prairie;—some level and wet land in the south-eastern part. Timber in sufficient quantities.

Cook.—Adjoins Lake Michigan, and has the branches of Chicago, Des Plaines, Du Page, Au Sable and Hickory creeks. Surface, tolerably level; rich soil,—extensive prairies,—timber in groves;—a few swamps. Plenty of limestone, and the streams run over rocky beds.

Crawford.—The Wabash river on its eastern side, with Lamotte, Hudson, Racoon and Sugar creeks. Some level prairies, rather sandy, with a full supply of timber.

Edgar.—Watered by Big, Clear, and Brulette's creeks on the eastern, and Little Embarras on its western side. Southern and eastern sides timbered; northern and western sides much prairie; some undulating,—some level and rather wet. Grand View is a delightful tract of country.

Edwards.—The Little Wabash on its western, and Bon Pas on its eastern border. Several prairies, high, undulating, and bounded by heavy timber. Soil, second quality.

Effingham.—Watered by the Little Wabash and its tributaries; due proportion of timber and prairie; tolerably level,—second rate.

Fayette.—Kaskaskia river, Hurricane, Higgens', Ramsey's and Beck's creeks. The bottom lands on the Kaskaskia low and inundated at high water; considerable prairie; much heavy timber; soil, second rate.

Franklin.—Watered by the Big Muddy and its branches, and the South Fork of Saline creek. The prairies small, fertile and level, —timber plenty,—soil rather sandy.

Fulton.—The Illinois on the south-eastern side, with Spoon river and several small creeks through it. About half heavily timbered, with rich, undulating prairies; streams flow over a pebbly bed; soil, first rate.

Gallatin.—Joins the Wabash and Ohio rivers, and has the Saline and branches running through it. Soil, sandy, with sandrock, limestone, quartz crystals, excellent salines, &c. Timber of various kinds; no prairies.

Greene—Has the Mississippi south, the Illinois west, with Otter, Macoupen and Apple creeks. Much excellent land, both timber and prairie, in due proportion, with abundance of lime, and sandstone, and coal.

Hamilton.—Watered by branches of the Saline and Little Wabash; a large proportion, timbered land; soil, second and third rate, with some swamp in the northern part. Sandstone and some lime.

Hancock.—Besides the Mississippi, it has a part of Bear, Crooked, and Camp creeks; large prairies; timber along the streams; rich, first rate land.

Henry—Has Rock river north, with Winnebago swamp, and its outlet or Green river, and one of the heads of Spoon river, and Edwards river interior. Some rich, undulating prairies and groves, with considerable wet, swampy land. Not much population.

Iroquois.—Kankakee, Iroquois and Sugar creek. Sand ridges and plains; much rich prairie; some timber, but deficient. It is found chiefly in groves and strips along the water courses.

Jackson—Has the Mississippi on the south-west, and Muddy river running diagonally through it, with some of its tributaries. Some prairies in the north-eastern part,—much heavy timber,—some hilly and broken land,—with abundance of coal, saline springs, lime and sandstone.

Jasper.—The Embarras runs through it, and the Muddy Fork of the Little Wabash waters its western side. Much of both the prairie and timbered land is level and rather wet; some fertile tracts.

Jefferson.—Watered by several branches of the Big Muddy and Little Wabash. Soil, second rate; surface, a little undulating; one third prairie; several sulphur and other medicinal springs.

Jo Daviess—Formerly embraced all the State north-west of Rock river, but recently divided into three or four counties. Besides the Mississippi, it has Fever river, Pekatonokee, Apple river, and Rush and Plum creeks. A rich county, both for agricultural and mining purposes. Timber scarce, and in groves; surface, undulating,—in some places hilly; well watered by streams and springs, and has good mill sites. Copper and lead ore in abundance.

Johnson.—The Ohio on the south, Cash river and Big Bay creek, and a series of lakes or ponds interior. A timbered country, tolerably level; soil sandy, with considerable quantities of second rate land.

Knox.—Watered by Henderson and Spoon rivers, and their tributaries. The prairies large, moderately undulating, and first quality of soil, with excellent timber along the water courses.

Lasalle.—Besides the Illinois river, which passes through it, Fox river, Big and Little Vermilion, Crow, Au Sable, Indian, Mason, Tomahawk, and other creeks, water this county. They generally run on a bed of sand or lime rock, and have but little alluvial bottom lands. Deficient in timber, but has an abundance of rich, undulating prairie, beautiful groves, abundant water privileges, and extensive coal banks.

Lawrence.—The Wabash east, Fox river west, and Embarras and Racoon through it. An equal proportion of timber and prairie, some excellent, other parts inferior,—and some bad, miry swamps, called "*purgatories.*"

Macon.—South-east portion, watered by the Kaskaskia and tributaries; the middle and northern portions by the North Fork of Sangamon, and the north-western part by Salt creek. The prairies large, and in their interior, level and wet,—towards the timber, dry, undulating and rich.

Madison.—The Mississippi lies west; Cahokia and Silver creeks, and Wood river, run through it. A part of this county lies in the American bottom, and is a rich and level alluvion; but much of the county is high, undulating, and proportionably divided into timber and prairie. Well supplied with stone quarries and coal banks.

Macoupen.—The Macoupen creek and branches water its central and western parts, the Cahokia, the south-eastern, and the heads of Wood river and Piasau, the south-western parts. A large proportion of the county is excellent soil, well proportioned into timber and prairie, and slightly undulating.

Marion.—Watered by the East Fork, and Crooked creek, tributaries of Kaskaskia river, on its western, and heads of Skillet Fork of Little Wabash on its eastern sides. Much of the land of second quality, slightly undulating, about one third timbered,—some of the prairie land level, and inclined to be wet.

McDonough.—Crooked creek and its branches water most of the county. The eastern side, for eight or ten miles in width, is prairie,—the western and middle parts suitably divided between prairie and forest land; surface, moderately undulating; soil, very rich.

McLean.—One third of the eastern, and a portion of the northern side, is one vast prairie. The timber is beautifully arranged in groves; the surface moderately undulating, and the soil dry and rich. The head waters of the Sangamon, Mackinau, and the Vermilion of the Illinois, are in this county. Its minerals are quarries of lime and sandstone, and granite bowlders, scattered over the prairies.

Mercer—Has the Mississippi on the west, and Pope and Edwards rivers interior, along which are fine tracts of timber; in its middle and eastern parts are extensive prairies; surface, generally undulating; soil, rich.

Monroe.—Watered by Horse, Prairie de Long, and Fountain creeks. The American bottom adjacent to the Mississippi is rich alluvion, and divided into timber and prairie. On the bluffs are ravines and sink-holes, with broken land. Further interior is a mixture of timber and prairie. Abundance of limestone, coal, and some copper.

Montgomery.—Watered by Shoal creek

and branches, and Hurricane Fork. Surface, high and undulating, and proportionably divided into timber and prairie. Soil, second rate.

Morgan.—A first rate county,—well proportioned into prairie and forest lands,—much of the surface undulating; watered by the Illinois river and Mauvaise-terre, Indian, Plum, Walnut, and Sandy creeks, and heads of Apple creek. Coal, lime and freestone.

Peoria.—Watered by the Illinois, Kickapoo, Copperas, Senatchwine, and heads of Spoon river. Surface, moderately rolling, rich soil, and proportionably divided into prairie and forest.

Perry.—Streams; Big Beaucoup, and Little Muddy; one third prairie, tolerably level, and second rate soil.

Pike.—-Besides Mississippi and Illinois, which wash two sides, it has the Suycartee slough running through its western border, and navigable for steam-boats, and a number of smaller creeks. The land and surface various,—much of it excellent undulating soil, —some rich alluvion, inundated at high water, —large tracts of table land, high, rolling, and rich, with due proportion of timber and prairie. A large salt spring.

Pope.—With the Ohio river east and south, it has Big Bay, Lusk's, and Big creeks interior. A timbered region, tolerably level, except at the bluffs, with good sandy soil, and sand and limestone.

Putnam.—The Illinois runs through it,—Spoon river waters its north-western part, and Bureau, Crow, Sandy, and some other streams, water its middle portions. Here are beautiful groves of timber, and rich, undulating and dry prairies, fine springs, and good mill sites. Lime, sand and freestone, and bituminous coal. A few tracts of wet prairie, with some ponds and swamps, are in the north-western part.

Randolph—Has the Mississippi along the western side; Kaskaskia river passes diagonally through it; soil, of every quality, from first rate to indifferent; surface, equally as various, with rocky precipices at the termination of the alluvial bottoms.

Rock Island—Is at the mouth of Rock river, which, with the Mississippi and some minor streams, drain the county. Rich alluvion along the Mississippi, with much excellent table land,—both timber and prairie interior. Some wet, level prairie, south of Rock river.

Sangamon.—Watered by Sangamon river and its numerous branches. Much of the soil is of the richest quality, with due proportions of timber and prairie, moderately undulating, and a first rate county.

Schuyler.—The south-eastern side has the Illinois, the interior has Crooked and Crane creeks, and the south-west has McKee's creek. Along the Illinois is much timber, with some inundated bottom lands. Interior, there is a due proportion of prairie, and timber, and rich soil, with an undulating surface.

Shelby—Is watered by the Kaskaskia and tributaries; has a large amount of excellent land, both timber and prairie, with good soil, moderately undulating.

St. Clair.—The streams are Cahokia, Prairie du Pont, Ogle's, Silver, Richland, and Prairie de Long creeks, and Kaskaskia river. The land is various, much of which is good, first and second rate, and proportionably divided into timber, prairie, and barrens. The minerals are lime and sandstone, and extensive beds of coal and shale.

Tazewell.—Watered by the Illinois, Mackinau, and their tributaries. Much of the surface is undulating, soil rich; prairie predominates, but considerable timber, with some broken land about the bluffs of Mackinau, and some sand ridges and swamps in the southern part of the county.

Union.—Watered by the Mississippi, Clear creek, the heads of Cash, and some of the small tributaries of the Big Muddy. Much of the surface is rolling and hilly,—all forest land. Soil, second and third rate. Some rich alluvial bottom.

Vermilion—Is watered by Big and Little Vermilion of the Wabash, with large bodies of excellent timber along the streams, and rich prairies interior. Surface, undulating and dry; soil, deep, rich and calcareous.

Wabash—Has Wabash river on the east,

Bon Pas on the west, and some small creeks central; surface, rolling, and a mixture of timber and prairie; soil, generally second rate. Minerals; lime and sandstone.

Warren.—Besides the Mississippi, its principal stream is Henderson river, which passes through it, with Ellison, Honey, and Camp creeks. Much of the land on these streams is rich, undulating, deficient somewhat in timber, with excellent prairie. Along the Mississippi, and about the mouth of Henderson, the land is inundated in high water.

Washington—Has the Kaskaskia on its north-western side, with Elkhorn, Little Muddy, Beaucoup, and Little Crooked creeks interior. The prairies are rather level, and in places inclined to be wet; the timber, especially along the Kaskaskia, heavy.

Wayne.—The Little Wabash, with its tributaries, Elm river, and Skillet Fork, are its streams. It is proportionably interspersed with prairie and woodland, generally of second quality.

White.—The eastern side washed by the Big Wabash, along which is a low, inundated bottom; the interior is watered by the Little Wabash and its tributaries. Some prairie, but mostly timber. Soil and surface various. Some rich bottom prairies, with sandy soil.

Towns. VANDALIA is the seat of government till 1840; after which, it is to be removed to Alton, according to a vote of the people, in 1834,—unless they should otherwise direct.

It is situated on the right bank of the Kaskaskia river, in north latitude 39° 0′ 42″, and fifty-eight miles in a direct line, a little north of east from Alton. The public buildings are temporary. Population, about seven hundred and fifty.

Alton. Two towns of this name are distinguished as Alton, and Upper Alton. Alton is an incorporated town, situated on the bank of the Mississippi, two and a half miles above the mouth of the Missouri, and at the place where the curve of the Mississippi penetrates the furthest into Illinois, eighteen miles below the mouth of the Illinois river. For situation, commerce, business of all kinds, health, and rapidity of growth, it far exceeds any other town on the east bank of the Mississippi, above New Orleans. The population is about two thousand. The commercial business done here is already immense, and extends through more than half of Illinois, besides a large trade on the western side of the Mississippi. Five large mercantile establishments do wholesale business only, four do wholesale and retail, besides four wholesale and retail groceries, and fifteen or twenty retail stores and groceries; and yet many more mercantile houses are necessary for the business of the country. Great facilities for business of almost every description, especially for every kind of mechanics, are to be had here. It offers one of the best situations on the western waters for building and repairing steam-boats. Town lots and lands adjacent, have risen in value from five hundred to one thousand per cent. within the last twelve months.

Alton has respectable and well finished houses of worship, for the Presbyterian, Methodist Protestant, and Baptist denominations; two good schools, a lyceum, that holds weekly meetings, and two printing-offices. The population in general, is a moral, industrious, enterprising class. Few towns in the West have equalled this in contributions for public and benevolent objects, in proportion to age and population.

Arrangements have been made for doing an extensive business in the slaughtering and packing of pork and beef. Four houses are engaged in that line, which slaughtered about twenty-five thousand hogs the last season. Many buildings will be erected the present season, amongst which will be an extensive hotel, which is much needed. The town is situated at the base, side, and top, of the first bluffs that extend to the river, above the mouth of the Kaskaskia. Adjacent to it, and which will eventually become amalgamated, is Middleton, laid off directly in the rear.

Upper Alton is from two and a half to three miles back from the river, and in the rear of Lower Alton, on elevated ground, and in every respect a very healthy situation. It has exceeding one hundred and twenty families, and is rapidly improving. Adjacent to it, and forming now a part of the town plat, is "*Shurtleff College, of Alton, Illinois,*" which bids fair to become an important and flourishing institution. Also, "*Alton Theological Seminary,*" which has commenced operations. Both these institutions have been gotten up under the influence and patronage of the Baptist denomination. A female seminary, of a high order, under the name of the "*Alton Female Institute,*" has been chartered, and a building is about to be erected for the purpose. The Baptists, Methodists and Presbyterians have congregations here, and two houses of worship were built the past year.

Chicago is the largest commercial town in Illinois. It is situated at the junction of north and south branches, and along the main Chicago, near its entrance into lake Michigan, on a level prairie, but elevated above the highest floods. A recent communication, from a respectable mercantile house, gives the following statistics: "Fifty-one stores, thirty groceries, ten taverns, twelve physicians, twenty-one attornies, and 4000 inhabitants. We have four churches, and two more building, one bank, a marine and fire insurance company about to go into operation, and a brick hotel containing ninety apartments.

There were nine arrivals and departures of steam-boats in 1835, and two hundred and sixty-seven, of brigs and schooners, containing 5015 tons of merchandise, and 9400 barrels of salt, besides lumber, provisions, &c.

The harbor now constructing by the United States government, will be so far completed in 1836, as to admit vessels and steam-boats navigating the lakes. A few miles back of Chicago, are extensive tracts of wet prairie.

Galena is the seat of justice for Jo Daviess county, situated on Fever river, in the midst of the mining district. It has about twenty stores, a dozen groceries, and about 1000 inhabitants.

Springfield is near the geographical centre of the State, and in the midst of a most fertile region of country. It is a flourishing inland town, and contains about 2000 inhabitants. *Jacksonville*, the county seat of Morgan county, has about the same population, and is equally delightful and flourishing.

One mile west, on a most beautiful eminence, stands "*Illinois College*," founded under the auspices of the Presbyterian denomination, and bids fair to become a flourishing seat of learning.

I have not room to name, much less describe, the many growing towns and villages in this State, that excite and deserve the attention of emigrants. On the Illinois river, are Ottawa, and several eligible sites in its vicinity, where towns have commenced; Beardstown, a short distance below the mouth of Sangamon river; Peoria, at the foot of Peoria lake (a most beautiful site, and containing 1,000 inhabitants); Meredosia, Naples, Pekin, Hennepin, &c. On the Mississippi, are Quincy, Warsaw, New Boston, and Stephenson, the seat of justice for Rock Island county. Interior, are Bloomington, Decatur, Tremont, Shelbyville, Hillsboro', Edwardsville, Carlyle, Belleville, Carrollton, and many others. Towards the Wabash, are Danville, Paris, Lawrenceville, Carmi, and Mount Carmel; the last of which has an importance, from being connected with the grand rapids of the Wabash. Shawneetown is the commercial depôt for the south-eastern part of the State. On the military tract, are Rushville, Pittsfield, Griggsville, Carthage, Macomb, Monmouth, Knoxville, Lewistown, Canton, &c.; all pleasant sites, and having a population from two or three hundred to one thousand inhabitants.

For a more particular description of each county, town and settlement, with all other particulars of Illinois, the reader is referred to "A Gazetteer of Illinois," by the author of this Guide.

Projected Improvements. The project of uniting the waters of lake Michigan and the Illinois, by a canal, was conceived soon after the commencement of the Grand canal of New York; and a board of commissioners, with engineers, explored the route and estimated the cost, in 1823. Provision, by a grant of each alternate section of land within five miles of the route, having been granted by Congress, another board of commissioners was appointed in 1829; a new survey was made, and the towns of Chicago and Ottawa laid off, and some lots sold in 1830. Various movements have since been made, but nothing effectually done, until the recent special session of the legislature, when an act was passed, to authorize the governor to borrow funds upon the faith of the State; a new board of commissioners has been organized, and this great work is about to be prosecuted with vigor to its completion.

Funds, in part, have been provided, from the sales of certain saline lands belonging to the State, to improve the navigation of the Great Wabash, at the Grand Rapids, near the mouth of White river, in conjunction with the State of Indiana. From the same source, funds are to be applied to the clearing out of several navigable water-courses, and repairing roads within the State.

Charters have been granted to several railroad companies, some of which have been surveyed, and the stock taken. One from Alton to Springfield was surveyed last year, and the stock subscribed in December. Another from St. Louis, by the coal mines of St. Clair county, to Belleville, thirteen miles, is expected to be made immediately. The project of a central rail-way from the termination of the Illinois and Michigan canal, at the foot of the rapids, a few miles below Ottawa, through Bloomington, Decatur, Shelbyville, Vandalia, and on to the mouth of the Ohio river, has been entered upon with spirit.

Another charter contemplates the continuance of a route, already provided for in Indiana, and noticed under Ohio, from Lafayette, Ia., by Danville, Shelbyville, and Hillsboro', to Alton, the nearest point from the east to the Mississippi. A rail-road charter was granted, at a previous session of the legislature, from Mercdosia to Jacksonville, and another from Vincennes to Chicago.

We have only room to mention the following charters, which have recently been granted, in addition to those already specified:

One from Pekin to Tremont, in Tazewell county, nine miles.

One from the Wabash, by Peoria, to War saw, in Hancock county.

The Wabash and Mississippi rail-road company.

The Mount Carmel and Alton rail-road company.

The Rushville rail-road company.

The Winchester, Lynville, and Jacksonville rail-road company.

The Shawneetown and Alton rail-road company.

The Pekin, Bloomington, and Wabash rail-road company.

The Waverly and Grand Prairie rail-road company.

The Galena and Chicago union rail-road company.

The Wabash and Mississippi union rail-road company.

The Mississippi, Carrollton, and Springfield rail-road company.

The *national road* is in progress through this State, and considerable has been made on that portion which lies between Vandalia and the boundary of Indiana. This road enters Illinois at the north-east corner of Clark county, and passes diagonally through Coles and Effingham counties, in a south-westerly course, to Vandalia, a distance of ninety miles. The road is established eighty feet wide, the central part thirty feet wide, raised above standing water, and not to exceed three degrees from a level. The base of all the abutments of bridges must be equal in thickness to one third of the height of the abutment.

The road is not yet placed in a traveling condition. The line of the road is nearly direct; the loss, in ninety miles, being only the eighty-eighth part of one per cent. Between Vandalia and Ewington, for twenty-three miles, it does not deviate in the least from a direct line.

From Vandalia, westward, the road is not yet located; but it will probably pass to Alton.

Education. The same provision has been made for this as other Western States, in the disposal of the public lands. The section, numbered sixteen, in each township of land, is sold upon petition of the people within the township, and the avails constitute a permanent fund, the interest of which is annually applied towards the expenses, in part, of the education of those who attend school, living within the township.

A school system, in part, has been arranged by the legislature. The peculiar and unequal division of the country into timber and prairie lands, and the inequality of settlements consequent thereupon, will prevent, for many years to come, the organization of school districts with *defined geographical boundaries.* To meet this inconvenience, the legislature has provided, that any number of persons can elect three trustees, employ a teacher in any mode they choose, and receive their proportion of the avails of the school funds. *In all cases, however, the teacher must keep a daily account of each scholar who attends school, and make out a schedule of the aggregate that each scholar attends, every six months*, and present it, certified by the trustees of the school, to the school commissioner of the county, who apportions the money accordingly.

This State receives three per cent. on all the net avails of public lands sold in this State, which, with the avails of two townships sold, makes a respectable and rapidly increasing fund, the interest only of which can be expended, and that only to the payment of instructers.

Good common school teachers, both male and female, are greatly needed, and will meet with ready employ and liberal wages. Here is a most delightful and inviting field for

Christian activity. Common school, with Sunday school instruction, calls for thousands of teachers in the West.

Several respectable academies are in operation, and the wants and feelings of the community call for many more. Besides the colleges at Jacksonville and Alton, already noticed, others are projected, and several have been chartered. The Methodist denomination have a building erected, and a preparatory school commenced, at Lebanon, St. Clair county. The Episcopalians are about establishing a college at Springfield. One or more will be demanded in the northern and eastern portions of the State; and it may be calculated, that, in a very brief period, the State of Illinois will furnish facilities for a useful and general education, equal to those in any part of the country.

Government. The constitution of Illinois was formed by a convention, held at Kaskaskia, in August, 1818. It provides for the distribution of the powers of government into three distinct departments,—the legislative, executive, and judiciary. The legislative authority is vested in a general assembly, consisting of a senate and house of representatives. Elections are held biennially, as are the ordinary sessions of the legislature. Senators are elected for four years.

The executive power is vested in the governor, who is chosen every fourth year, by the electors for representatives; but the same person is ineligible for the next succeeding four years. The lieutenant governor is also chosen every four years.

The judicial power is vested in a supreme court, and such inferior courts as the general assembly, from time to time, shall establish. The supreme court consists of a chief justice and three associate judges.

The governor and judges of the supreme court constitute a council of revision, to which all bills, that have passed the assembly, must be submitted. If objected to by the council of revision, the same may become a law, by the vote of a majority of all the members elected to both houses.

The right of suffrage is universal. All white male inhabitants, citizens of the United States, twenty-one years of age, and who have resided within the State six months next preceding the elections, enjoy the right of voting. Votes are given *viva voce.* The introduction of slavery is prohibited. The constitution can only be altered by a convention.

General Remarks. 1. Farms, somewhat improved, are almost daily exchanging owners, and a considerable spirit of enterprise has been awakened within a year or two past. The prices of farms and improvements vary greatly, and are influenced much by factitious and local circumstances. From St. Clair county, northward, they average, probably, from five to ten dollars per acre, and are rising in value. In some counties, farms will cost from two to five dollars per acre. A *farm* in Illinois, however, means a tract of land, much of it in a state of nature, with some cheap, and, frequently, log-buildings; with twenty, forty, sixty, eighty or one hundred acres, fenced and cultivated. Good dwellings of brick, stone or wood, begin to be erected. Among the old residents, there have been but few barns built.

The want of adequate supplies of lumber, and of mechanics, renders good buildings more expensive than in the new counties of New England or New York.

2. Merchants' goods, groceries, household furniture and almost every necessary and comfort in house-keeping, can be purchased here; and many articles retail at about the same prices as in the Atlantic States.

3. The following table will exhibit the cost of three hundred and twenty acres of land, at Congress price, and preparing one hundred and sixty acres for cultivation or prairie land:

Cost of 320 acres, at $1 25 per acre, ...	$400
Breaking up 160 acres prairie, at $2 per acre, .	320
Fencing it into four fields, with a Kentucky fence of eight rails high, with cross stakes, ...	175
Cabins, corn-cribs, stable, &c., ...	250
Making the cost of the farm,	$1145

In many instances, a single crop of wheat will pay for the land, for fencing, breaking up, cultivating, harvesting, threshing, and taking to market.

4. All kinds of mechanical labor, especially

in the building line, are in great demand; even very coarse and common workmen get almost any price they ask. Journeymen mechanics get $2 per day. A carpenter or brick mason wants no other capital, to do first-rate business, and soon become independent, than a set of tools, and habits of industry, sobriety, economy and enterprise.

5. Common laborers on a farm obtain from $12 to $15 per month, including board. Any young man, with industrious habits, can begin here without a dollar, and in a very few years become a substantial farmer. A good cradler in the harvest-field, will earn from $1 50 to $2 per day.

6. Much that we have stated in reference to Illinois, will equally apply to Missouri, or any other Western State. Many general principles have been laid down, and particular facts exhibited, with respect to the general description of the State, soil, timber, kinds of land, and other characteristics, under Illinois, and, to save repetition, are omitted elsewhere.

CHAPTER XIII.

MISSOURI.

Extent and Boundaries—Civil Divisions—Population at different Periods—Surface, Soil and Productions—Towns.

Length, two hundred and seventy-eight; medium breadth, two hundred and thirty-five miles: containing 64,500 square miles, and 41,280,000 acres.

Bounded north by the Des Moines country, or New Purchase, attached to Wisconsin Territory, west by Indian Territory, south by Arkansas, and east by the Mississippi river;—Between 36° and 40° 37′ north latitude, and between 11° 15′ and 17° 30′ west longitude.

Civil Divisions. It is divided into fifty counties, as follows:—Barry, Benton, Boone, Callaway, Cape Girardeau, Carroll, Chariton, Clay, Clinton, Cole, Cooper, Crawford, Franklin, Gasconade, Green, Howard, Jackson, Jefferson, Johnson, Lafayette, Lewis, Lincoln, Madison, Marion, Monroe, Montgomery, Morgan, New Madrid, Perry, Pettis, Pike, Polk, Pulaski, Randolph, Ralls, Ray, Ripley, Rives, St. Francis, Ste. Genevieve, St. Charles, St. Louis, Saline, Scott, Shelby, Stoddart, Van Buren, Warren, Washington and Wayne.

Population at Different Periods.

In	Population.	From	Increase.
1810 (excluding Arkansas), ...	19,833	1810 to 1820, ..	46,753
1820, ...	66,586	1820 " 1824, ..	14,500
1824, ...	80,000	1824 " 1830, ..	60,455
1830, ...	140,455	1830 " 1832, ..	35,820
1832, ...	176,276	1832 " 1836, ..	33,724
1836, estimated, .	210,000		

The constitution is similar to that of Illinois, in its broad features, excepting the holding of slaves is allowed, and the General Assembly has no power to pass laws for the emancipation of slaves, without the consent of their owners, or paying an equivalent. It is made the duty of the General Assembly "to oblige the owners of slaves to treat them with humanity, and to abstain from all injuries to them extending to life or limb." "Slaves shall not be deprived of an impartial trial by jury." In 1832, there were in the State, 32,184 slaves, and 661 free colored persons. Every free white male citizen has the right of suffrage, after a residence in the State of one year.

Surface, Soil and Productions. The surface of this State is greatly diversified. South of

14*

Cape Girardeau, with the exception of some bluffs along the Mississippi, it is entirely alluvial, and a large proportion consists of swamp and inundated lands, the most of which are heavily timbered. From thence to the Missouri river, and westward to the dividing grounds between the waters of the Osage and Gasconade rivers, the country is generally timbered, rolling, and in some parts, quite hilly. No part of Missouri, however, is strictly mountainous. Along the waters of Gasconade and Black rivers the hills are frequently abrupt and rocky, with strips of rich alluvion along the water-courses. Much of this region abounds with minerals of various descriptions. Lead, iron, coal, gypsum, manganese, zinc, antimony, cobalt, ochre of various kinds, common salt, nitre, plumbago, porphyry, jasper, chalcydony, buhrstone, marble and freestone, of various qualities. The lead and iron ore are literally exhaustless, and of the richest quality. To say there is probably iron ore enough in this region to supply the United States with iron for one hundred thousand years to come, would not be extravagant. Here, too, is water power in abundance, rapid streams, with pebbly beds, forests of timber and exhaustless beds of bituminous coal. The only difficulty of working this vast body of minerals is the inconvenience of getting its proceeds to the Mississippi. The streams that rise in this region, run different courses into the Missouri, the Mississippi and the Arkansas, but they are too rapid and winding in their courses to afford safe and easy navigation.

Were the rafts now lodged in the St. Francois removed by the agency of government, as they have been in Red river, the lower section of the mineral country could be reached by steam-boat navigation. The citizens of St. Louis, very recently, have entered upon the project of a railway from that city, through the heart of this country, to the fine, farming lands, in the south-western part of the State. Such a project, carried into effect, would open a boundless field of wealth in Missouri.

The western part of the State is divided into prairie and forest land, much of which is fertile. Along the Osage, it is hilly, and the whole is undulating, and regarded as a healthy region, abounding with good water, salt springs and limestone. North of the Missouri the face of the country is diversified, with a mixture of timber and prairie. From the Missouri to Salt river, good springs are scarce, and in several counties it is difficult to obtain permanent water by digging wells. Artificial wells, as they may be called, are made by digging a well forty or fifty feet deep, and replenishing it with a current of rain water from the roof of the dwelling-house. Much of the prairie land in this part of the State is inferior to the first quality of prairie land in Illinois, as the soil is more clayey, and does not so readily absorb the water. Between Salt river and Des Moines, are beautiful, rich lands. The counties of Ralls, Marion, Monroe, Lewis and Shelby are first rate. The counties of Warren, Montgomery, Callaway, Boone, Howard and Chariton, all lying on the north side of the Missouri river, are rolling; in some places are bluffs and hills, with considerable good prairie, and an abundance of timbered land. Further west, the proportion of prairie increases to the boundary line, as it does to the northward of Boone, Howard and Chariton counties. After making ample deductions for inferior soil, ranges of barren hills, and large tracts of swamp, as in the south, the State of Missouri contains a vast proportion of excellent farming land. The people, generally, are enterprising, hardy and industrious; and most of those who hold slaves, perform labor with them. Emigrants from every State and several countries of Europe, are found here, but the basis of the population is from Kentucky, Tennessee and Virginia. The natural productions of Missouri are similar to those States already described, and the agricultural productions are the same as Illinois, except that more tobacco is produced in the middle, and considerable quantities of cotton in the southern counties.

Towns. The city of Jefferson is the political capital of the State. It is situated on the right bank of the Missouri, a few miles above the mouth of the Osage, and about one hundred and thirty-eight miles from St. Louis. It is a small town, with little business, except what pertains to the government of the State. A state house, governor's house and penitentiary have been erected.

St. Louis is the commercial capital, and the most important place in all this portion of the Valley of the Mississippi. It stands on the western bank of the Mississippi, one hundred and eighty miles above the junction of the Ohio, eighteen miles below that of the Missouri, and thirty-eight miles below that of the Illinois. It is beautifully situated on ascending and elevated ground, which spreads out into an undulating surface to the west for many miles. Two streets are parallel with the river on the first bank, and the rest of the city stands on the second bank; but very little grading is necessary, to give the streets running back from the river their proper inclination. The old streets, designed only for a French village, are too narrow for public convenience, but a large part of the city has been laid out on a liberal scale. The Indian and Spanish trade, the fur and peltry business, lead, government agencies, army supplies, surveys of government lands, with the regular trade of an extensive interior country, makes St. Louis a place of great business, in proportion to its population, which is about 10,000.

The following, from the register of the wharf-master, will exhibit the commerce for 1835:—

Steam-boat Register.

Number of different boats arrived,	121
Aggregate of tonnage,	15,470
Number of arrivals,	803
Wharfage collected,	$4,573 60

Wood and Lumber, liable to wharfage.

Plank, joist and scantling,	1,414,330 feet.
Shingles,	148,000 "
Cedar posts,	7,706 "
Fire-wood,	8,066 cords.

The proportionate increase of business will be seen by reference to the following registry for 1831:—

Different steam-boats arrived,	60
Average amount of tonnage,	7,769
Number of entries,	532

The morality, intelligence and enterprise of this city is equal to any other in the West, in proportion to its size. The American population is most numerous, but there are many French, Irish and Germans. About one third of the inhabitants are Roman Catholics. The Presbyterians, Methodists and Episcopalians have large congregations and houses of worship: the Baptists and Unitarians are rather small, and without public edifices. The Roman Catholic cathedral is a costly building, of freestone, and has a splendid chime of bells, sent over from Europe. St. Louis is a pleasant and healthy situation, and surrounded by a fertile country.

We have not space to give particulars respecting many interesting and flourishing towns in Missouri.

Cape Girardeau is a commercial depôt for the southern part of the State. Ste. Genevieve stands a little back from the river, and is known only as an old French village.

Selma is a landing and depôt for the lead mine country, thirty-eight miles below St Louis. Clarksville, Louisiana, Marion city, Hannibal, Saverton and La Grange are commercial sites on the Mississippi, above the mouth of Missouri. Palmyra is a beautiful town, of about one thousand inhabitants, and the seat of justice for Marion county. Along the Missouri, are Portland, Rocheport, Boonville, Lexington, Independence, and many other places of various degrees of importance. Franklin formerly stood on the north bank of Missouri, but most of it has been removed, three miles interior, to the bluffs. Potosi is a central town in the mineral district. Fulton, Columbia and Fayette are the seats of justice for Callaway, Boone and Howard counties, and are pleasant and flourishing towns.

About the same provision for education has been made in this as in other Western States, and a disposition to encourage schools, academies and colleges is fast increasing.

CHAPTER XIV.

ARKANSAS.

Situation and Extent—Civil Divisions—Rivers—Face of the Country—Soil—Water—Productions—Climate—Minerals—State of Society.

ARKANSAS, which has recently formed a constitution, and has been received into the national Union, lies between 33° and 36° 30′ north latitude, and between 13° 30′ and 17° 45′ west longitude. Length, two hundred and thirty-five; medium breadth, two hundred and twenty-two miles; containing about 50,000 square miles, or 32,000,000 acres.

Civil Divisions. The following are the counties, with the population, from the census taken in 1835:—

Counties.	Popula.	Counties.	Popula.	Counties.	Popula.
Arkansas, .	2,080	Independence,	2,633	Phillips, . . .	1,518
Carroll, . .	1,357	Izard, ...	1,879	Pike, ...	449
Chicot, . . .	2,471	Jackson, ...	891	Pope, ...	1,318
Conway, .	1,214	Jefferson, ...	1,474	Pulaski, . . .	3,513
Clark, . . .	1,285	Johnson, ...	1,803	Scott, ...	100
Crawford, .	3,139	Lafayette, . . .	1,446	Sevier, ...	1,350
Crittenden,	1,407	Lawrence, . . .	3,844	St. Francis, .	1,896
Greene, . . .	971	Miller, ...	1,373	Union, ...	878
Hempstead,	2,955	Mississippi, . .	600	Van Buren, .	855
Hot-Spring,	6,117	Monroe, ...	556	Washington,	6,742
				Total,	58,212

Another table we have seen, makes out the population, as officially reported (with the exception of two counties, from which returns had not been made), to be 51,809;—white males, 22,585; white females, 19,386;—total whites, 41,971: slaves, 9,629;—free persons of color, 209. The population, in 1830, 30,388;—in 1833, 40,660.

The following graphic description of Arkansas, is taken from a letter from Rev. Harvey Woods, in that State, to the editor of the Cincinnati Journal, and is corroborated by testimony in our possession, from various correspondents. It was written in 1835.

"Arkansas Territory is a part of that vast country ceded to the United States, by France, in 1803. From the time of the purchase, till lately, the tide of emigration hardly reached thus far. In 1800, the population was 1052. Arkansas was erected into a Territory, in 1819. At this time it is receiving a share of those who retire beyond the Mississippi.

"*Rivers.* The Territory is admirably intersected with navigable rivers; the Mississippi on the east, the great Red river on the south. Between these, and running generally from north-west to south-east, are the St. Francois, White, Arkansas and Washitau rivers, all fine streams for steam-boat navigation.

"*Face of the Country.* It is various. No country affords more diversified scenery. The country in the east, for one hundred miles, is flat, with marshes and swamps; in the middle, broken and hilly; and in the west, hilly and mountainous. There are some prairies, some thickly timbered land, some heavy timbered. The country is generally a timbered country. Some parts are sandy, some rocky, and some flinty.

" *Soil.* Should a man travel here, and expect to find all good land, he would be sadly disappointed. The best lands are generally contiguous to the rivers and creeks; and these are exceedingly fertile, not surpassed by any soil in the United States. Arkansas soil, that is rich, has just sand enough to make it lively and elastic. Our best lands are covered with walnut, hackberry, mulberry, oak, ash, grape vines, &c.

" *Water.* The hilly and mountainous parts are well supplied with springs, limestone and freestone. Also good streams for mills. In the flat country, good water is easily obtained by digging.

" *Productions.* Cotton and corn are the principal. The Arkansas cottons commanded the best price last season, in the Liverpool market. It is a country of unequalled advantages for raising horses, mules, cattle and hogs.

" *Climate.* It is mild, and, from its difference in latitude, say from 32° 40′ to 36° 30′ N., and the difference in local situation, we would guess, and correctly too, that there is much difference in the health of different places; the high and northern parts healthy, and the flat and southern subject to agues and bilious fevers. The climate has been considered unhealthy to new settlers; but it is not more so than other new countries.

"*Minerals.* There are quantities of iron, lead, coal, salt, and, it is asserted by some, silver. There are many salt and sulphur springs. On the Arkansas river, beyond the limits of the Territory proper, is a section of country called the salt prairie, which, according to good authority, is covered, for many miles, from four to six inches deep, with pure white salt. In the Hot Spring country, are the famous hot springs, much resorted to by persons of chronic and paralytic diseases. The temperature, in dry, hot weather, is at boiling point.

"*State of Society.* The general character of the people is brave, hardy and enterprising, —frequently without the polish of literature, yet kind and hospitable. The people are now rapidly improving in morals and intellect. They are as ready to encourage schools, the preaching of the gospel, and the benevolent enterprises of the age, as any people in new countries. The consequences of living here a long time, without the opportunity of educating their children, and destitute of the means of grace, are, among this population, just what they always will be under similar circumstances. Ministers of all denominations are 'few and far between.' We have no need *here* to build on others' foundation.

"I am living in Jackson county, on White river. This county has a larger quantity of good land than any one in the Territory. White river is always navigable for steamboats to this place, three hundred and fifty miles from its mouth. Well water is good,—some fine springs. Washington county, and some others, that have the reputation of better health, are more populous.

"We want settlers; and we have no doubt, that vast numbers of families in the States, particularly the poor, and those in moderate circumstances, would better their situation by coming here, where they can get plenty of fertile and fresh land at government price,—$1,25 per acre. They can have good range, and all the advantages of new countries. Emigrants, however, ought not to suffer themselves to expect all sunshine, and no winter. We have cloudy days and cold weather, even in Arkansas! If they have heard of the *honey pond*, where flitters grow on trees, they need not be surprised, if they do n't find it. Cabins cannot be built, wells dug, farms opened, rails made, and meeting-houses and school-houses erected, without work.

"It may be asked, 'If Arkansas be so fine a country, why has it not been settled faster?' There are, perhaps, three reasons:—a fear of the Indians, a fear of sickness, a fear of bad roads. The Indians are now all peaceably situated beyond the Territory proper, and are blessed with the labors of a number of good pious missionaries, who are teaching them to read the Bible, and showing the tall sons of the forest the way that leads to heaven. Sickness is no more to be dreaded here, than in Illinois and Missouri. The roads have indeed been bad. For a long time, no one could venture through the Mississippi swamps, unless he was a Daniel Boone. But appropriations have been made, by Congress, for several roads. This summer (1835), roads from Memphis to Little Rock, and to Litchfield and Batesville, and other points, will be completed. An appropriation of upwards of $100,000 has been made, to construct a road through the Mississippi swamp.

"Again: we want settlers,—we want physicians, lawyers, ministers, mechanics and farmers. We want such, however, and *only such*, as will make good neighbors. If any, who think of coming to live with us, are gamblers, drunkards, Sabbath-breakers, profane swearers, or the like, we hope that, when they leave their *old* country, they will leave their *old* habits."

The constitution of this State, in its essential features, is similar to Missouri, and other south-western States.

CHAPTER XV.

WISCONSIN TERRITORY.

Boundaries and Extent—Rivers—Soil—Productions—Towns, &c.

UNDER this name, is now comprehended an extensive district of country, lying on both sides of the Mississippi river, above Illinois and Missouri, and extending indefinitely north. That portion lying betwixt the northern boundary of Illinois and the Wisconsin river, and from lake Michigan to the Mississippi, has the Indian title extinguished, and, in part, has been surveyed and brought into market. There is much excellent land in this part of the Territory, and it is well watered with perennial streams and springs. Offices are opened, for the sale of public lands, at Mineral Point and Green Bay, and a large amount has been sold, and some at a high price. The country immediately bordering on lake Michigan, is well timbered, with various trees. Here are red, white, black and burr oaks, beech, ash, linden, poplar, walnut, hickory, sugar and white maple, elm, birch, hemlock and pine, with many other kinds. The soil is not so deep and dark a mould as in the prairies of Illinois, but is fertile and easily cultivated; it is sandy, especially about the town of Green Bay. Towards the lake, and near the body of water called Sturgeon Bay, connected with Green Bay, and between that and the lake, are extensive swamps and cranberry marshes. Wild rice, tamarisk and spruce grow here. About Rock river, and from thence to the Mississippi, there is much excellent land, but a deficiency of timber. Lead and copper ore, and probably other minerals, abound in this part of the country. Along to the east and north of the Four lakes, are alternate quagmires and sand-ridges, for fifty miles or more, called by the French *coureurs du bois*, "*terre tremblant*" (trembling land), the character of which is sufficiently indicated by the name.

There are several small lakes in the district of country we are now examining; the largest of which is Winnebago, situated thirty or forty miles south of Green Bay. It is about ten miles long and three broad, and is full of wild rice: Fox river passes through it. Kushkanong is situated on Rock river, between Catfish and Whitewater; it is six or eight miles in diameter, with some swamps and quagmires in its vicinity.

The *Four lakes* are strung along on a stream called Catfish, which enters Rock river twenty-five or thirty miles above the boundary of Illinois. They are six or eight miles long, abounding with fish, and are surrounded with an excellent farming country.

Green Bay settlement and village is two hundred and thirty miles north of Chicago; two hundred and twenty north-east from Galena; one hundred and twenty from Fort Winnebago, and in north latitude 44° 44′. *Navarino* is a town recently commenced in this vicinity, with an excellent harbor, grows rapidly, and bids fair to become a place of importance. Property has risen the last year most astonishingly.

Fort Winnebago is a military post, at the bend, and on the right bank of Fox river, opposite the portage. From thence to the Wisconsin, is a low, wet prairie, of three fourths of a mile, through which, a company has been chartered to cut a canal. On this route, the first explorers reached the Mississippi, in 1673. The Wisconsin river, however, without considerable improvement, is not navigable for steam-boats, at ordinary stages of the water, without much trouble. It is full of bars, islands, rocks, and has a devious channel.

The streams that rise in the eastern part of this Territory and flow into lake Michigan, north of the boundary of Illinois, are in order as follows: 1. Pipe creek, a small stream, but a few miles from the boundary: 2. Root river: 3. Milwaukee, ninety miles from Chicago, which rises in the swampy country, south of Winnebago lake, runs a south-easterly course, and after receiving the Menomone, forms Milwaukee bay. Here is a town site, on both sides of the river, with a population of six or eight hundred, which promises to become a place of business. The soil up the Milwaukee is good, from six to twelve inches in depth, a black loam and sand. Passing northward down the lake, is Oak creek, nine miles below Milwaukee; thence twenty-one miles is Sauk creek, a small stream. Seventy miles from Milwaukee is Shab-wi-wi-a-gun. Here is found white pine, maple, beech, birch and spruce, but very little oak: the surface level and sandy. Pigeon river is fifteen or twenty miles further on, with excellent land on its borders; timber, maple, ash, beech, linden, elm, &c. Fifteen miles further down is Manatawok. Here commences the hemlock, with considerable pine. This stream is about forty or fifty miles from Green Bay settlement. Twin rivers are below Manatawok, with a sandy soil, and good timber of pine and other varieties. From Milwaukee to Green Bay, by a surveyed route, is one hundred and twelve miles; by the Indian trail, commonly traveled, one hundred and thirty-five miles. North of the Wisconsin river is Crawford county, of which Prairie du Chien is the seat of justice. From the great bend at Fort Winnebago, across towards the Mississippi, is a series of abrupt hills, rising several hundred feet, and covered with a dense forest of elm, linden, oak, walnut, ash, sugar-maple, &c. The soil is rich, but too hilly and broken for agricultural purposes. There is no alluvial soil or bottoms along the streams, or grass in the forests.

The Wisconsin river rises in an unexplored country, towards lake Superior. The *coureurs du bois* and *voyageurs* represent it as a cold, mountainous, dreary region, with swamps.

West of the Mississippi, above Des Moines, and extending northward to a point some distance above the northern boundary of Illinois, and for fifty miles interior, is a valuable country, purchased of the Indians, in 1832. Its streams rise in the great prairies, and run an east or south-east course into the Mississippi. The most noted are Flint, Skunk, Wau-be-se-pin-e-con, Upper and Lower Iowa rivers, and Turkey, Catfish and Big and Little Ma-quo-ka-tois or Bear creeks. The soil, in general, is excellent, and very much resembles the military tract in Illinois. The water is excellent, plenty of lime, sand and freestone, extensive prairies, but a deficiency of timber a few miles interior. About Dubuque, opposite Galena, are extensive and rich lead mines. Burlington is a town at the Flint hills, opposite Warren county, Illinois, containing a population of seven hundred. Dubuque is situated on the Mississippi, on a sandy bottom, above high water, and fourteen miles north-west from Galena. It has about sixty stores and groceries, two taverns, two churches, and about one thousand inhabitants, and we have before us the prospectus for the "*Dubuque Visitor*," a weekly newspaper. Peru is in the vicinity, and contains about five hundred inhabitants. The New Purchase, as this district of country is called, is divided into two counties, Dubuque and Des Moines, and contains a population of eight or ten thousand.

This Territory has, heretofore, for civil purposes, formed a part of the late Michigan Territory;—but in 1836 it was erected into a territorial government, by an act of Congress. The population has been recently estimated, by the legislature of the Territory, at 30,000.

Probably not many years will elapse before two new States will be formed out of this district of country, the one on the eastern and the other on the western side of the Mississippi.

CHAPTER XVI.

LITERARY AND RELIGIOUS INSTITUTIONS.

Colleges—Statistical Sketch of each Denomination—Field for Effort, and Progress made.

In giving a sketch of literary and religious institutions in the West, the very limited space remaining to be occupied in this work, compels me to throw together a few general facts only. The author has made some progress in collecting materials, and he designs to prepare another work soon, in which a variety of particulars and sketches will be given of the early history, progress of literary and religious institutions, colleges, seminaries and churches, Bible, Sunday school, education, and other kindred societies in the Western Valley, with the present aspect of each denomination of Christians. The interest taken in the affairs of the West, and the anxiety evinced by the community for facts and particulars on those subjects, demand that they should be treated more in detail than the limits of this Guide will allow.

I. COLLEGES.

WESTERN PENNSYLVANIA—Has *Jefferson College*, at Canonsburgh, which the Presbyterians originated in 1802, from the first grammar school ever established by Protestants west of the Alleghany mountains. Graduates, in 1835, forty-six; new students admitted, seventy-five; present number, two hundred and thirty (including the preparatory department), of which one hundred and thirty-five profess religion. Course of mathematics and physical sciences greatly extended, with practical application to civil engineering. Instruction provided in Hebrew, French, Spanish, German, Portuguese and Italian;—provision for manual labor;—expenses moderate.

Washington College, at Washington, Penn., also connected with the Presbyterian denomination, founded in 1806; had one hundred and forty students in 1832.

Alleghany College, at Meadville, was founded in 1815, by Rev. T. Alden, has a valuable library of eight thousand volumes, principally the donation of the late Rev. Dr. *Bentley*, of Salem, Mass., a distinguished benefactor of this institution. The college did not flourish for some years, and it is now transferred to the Methodist Episcopal church, and is under charge of the Pittsburgh Conference. It now promises to be successful.

The *Western University of Pennsylvania* was founded in 1820. The number of graduates, in 1832, was fifty; of under graduates, in all departments, seventy. A beautiful college edifice has been erected in the western part of Pittsburgh, for this institution.

There is no collegiate institution in Western Virginia.

OHIO.—*Ohio University*, at Athens, was founded in 1802; has an endowment of forty-six thousand and eighty acres of land, which yields $2,300 annually. A large and elegant edifice of brick was erected in 1817. The number of students, about ninety.

Miami University was founded in 1824, and is a flourishing institution at Oxford, Butler county, thirty-seven miles from Cincinnati. It possesses the township of land in which it is situated, and from which it receives an income of about $5000. Number of students, about two hundred. Patronized by Presbyterians.

The *Cincinnati College* was incorporated in 1819,—continued to be sustained as a classical institution for some years, and then suspended operations. It has been revived and reörganized lately and will probably be sustained.

Kenyon College, at Gambier, Knox county, in a central part of the State, was established in 1828, through the efforts of Rev. Philander Chase, then bishop of the Ohio Diocess, who obtained about $30,000 in England to endow it. Its chief patrons were those excellent British noblemen, lords Kenyon and Gambier. It is under Episcopal jurisdiction, and has a theological department, for the education of candidates for the ministry in the Episcopal church. It has about one hundred and fifty students.

Western Reserve College is at Hudson. It was founded by Presbyterians and Congregationalists, in 1826, and has eighty-two students in all its departments.

Franklin College is in New Athens, Harrison county, on the eastern side of the State, and has about fifty students.

The *Granville Literary and Theological Institution* originated under patronage of the Baptist denomination in 1831. It is designed to embrace four departments,—preparatory, English, collegiate, and theological. It is rapidly rising, and contains more than one hundred students.

Oberlin Institute has been recently established in Lorrian county, under the influence of "new measure" Presbyterians, with four departments, and has two hundred and seventy-six students, as follows: In the theological department, thirty-five; collegiate, thirty-seven; preparatory, thirty-one; female, seventy-three. The citizens of Cleaveland have recently contributed to it $15,000, of which six persons gave $1000 each.

The *Willibough Collegiate Institute* is in the lake country of Ohio, and has been gotten up within a few years past.

Marietta Collegiate Institute is said to be a flourishing and respectable institution, having a large number of students in various departments.

INDIANA.—*Indiana College* is a State institution, established at Bloomington, and commenced operations in 1828. Present number of students not known. In 1832, the number exceeded fifty.

Hanover College is at South Hanover, six miles below the town of Madison, and near the Ohio river. It is a flourishing institution, with arrangements for manual labor, and is styled "South Hanover College and Indiana Theological Seminary." The number of students exceeds one hundred.

Wabash College, at Crawfordsville, has just commenced operations under auspicious circumstances. Under patronage of the Presbyterians.

ILLINOIS.—*Illinois College*, near Jacksonville, commenced as a preparatory school in 1830, and has made rapid progress. Large funds for its endowment have been recently provided in the Eastern States. The number of students about eighty.

Shurtleff College of Alton, Illinois, was commenced under the efforts of Baptists at Alton, in 1832, as a preparatory institution;—chartered, as a college, in February, 1835, and has been recently named in honor of a liberal patron, Dr. Benjamin Shurtleff, of Boston, Mass., who has presented the institution with $10,000. It has sixty students, and its prospects are encouraging.

McKendreean College has been chartered, a building erected, and a school commenced at Lebanon. It is connected with the Methodist Episcopal church. Charters have been recently granted for other colleges in this State, and measures adopted to bring some of them into existence. The Rev. Philander Chase, whose persevering labors brought into existence and successful operation, Kenyon college in Ohio, and who is now bishop of Illinois, is at present in England, where, by recent advices, he has obtained $50,000 to invest in Illinois lands, and to establish a college for the interests of the Episcopal church.

MISSOURI. — The Roman Catholics have two institutions of a collegiate character, established in this State.

St. Mary's College, in Perry county, was established by bishop Du Bourg, in 1822. It has six thousand volumes in the library. Including the *nunnery*, and school for females, a seminary for the education of *priests*, a preparatory, and a primary school, the number of teachers and students, are about three hundred.

St. Louis University was founded in 1829, and is conducted by the Fathers of the Society of Jesuits. The edifice is one hundred and thirty feet by forty, of four stories, including the basement, and is situated on elevated and pleasant ground, on the confines of the city.

For the Protestants, the following institutions have been established:

Columbia College, adjacent to Columbia, Boon county. The institution opened in 1835, under encouraging circumstances.

Marion College is in a delightful tract of country, a prairie region, in the western part of Marion county,—and has between eighty and one hundred students. It is connected with the Presbyterian interests. The project, as developed by some of its founders, is an immense one, including English, scientific, classical, theological, medical, agricultural, and law departments,—all to be sustained by manual labor, and the proceeds of extensive farms. Doubtless, by prudent and persevering efforts, a respectable college may be brought into successful operation.

A *college* at St. Charles, has been founded, principally by the liberality of George Collier, a merchant of St. Louis, and two or three other gentlemen, and a classical and scientific school has been commenced.

ARKANSAS.—Efforts are making to establish a college by Presbyterian agency, at Cane Hill, in this newly formed State. Two or three collegiate institutions will soon be needed in this region.

KENTUCKY. — *Transylvania University* at Lexington, is the oldest collegiate institution in the West. It was commenced, by a grant of eight thousand acres of land by the legislature of Virginia, in 1783, and was then called "Transylvania Seminary." The "Kentucky Academy" was founded in 1794, and both institutions were united and incorporated in 1798, under the present name. It has classical, medical, law, and preparatory departments, and including each, from three to four hundred students.

Center College, at Danville, was founded by the Presbyterian church, in 1818, for which the synod of Kentucky pledged $20,000. Number of students, about one hundred.

Augusta College was founded in 1822, by the Ohio and Kentucky conferences of the Methodist Episcopal church. It adopted collegiate regulations in 1828. Number of students in the collegiate, academical and primary departments, about two hundred.

Cumberland College was incorporated in 1824, and is established at Princeton, in the western part of the State. It is under the patronage and jurisdiction of the Cumberland Presbyterians. A farm, including a tract of five thousand acres of land, with workshops, furnish facilities for manual labor. It has about eighty students.

St. Joseph's College, is a Roman Catholic institution, at Bardstown, with college buildings sufficient to accommodate two hundred students, and valued at $60,000. It commenced with four students in 1820. In 1833, there were in the collegiate and preparatory departments, one hundred and twenty students. The St. Thomas and St. Mary seminaries, are also under the charge of Roman Catholic priests, the one in Nelson county, four miles from Bardstown, and the other in Washington county.

Georgetown College, in Scott county, was founded by the Baptist denomination, in 1830; but for some years it has been in other hands, and in not a prosperous condition. The Campbellites obtained a preponderating influence over it. After much contention, an arrangement has been effected to place it under the influence of the denomination to whom it fairly belonged, and the Rev. B. F. Farnsworth has been elected president, and has accepted the office. It is expected to prosper under this arrangement.

TENNESSEE.—The *University of Nashville* is a prominent institution. The laboratory is one of the finest in the United States, and the mineralogical cabinet not exceeded, and this department, as well as every other in the college, is superintended with much talent. The number of students is about one hundred.

Greenville, *Knoxville* and *Washington* colleges are in East Tennessee.

Jackson College is about to be removed from its present site, and located at Columbia; $25,000 have been subscribed for the purpose. A Presbyterian Theological Seminary is at Maryville.

MISSISSIPPI.—*Jefferson College* is at Washington, six miles from Natchez. It has not flourished as a college, and is now said to be conducted somewhat on the principle of a military academy.

Oakland College has been recently founded by Presbyterians, and bids fair to exert a beneficial influence upon religion and morals, much needed in that State. The Baptist denomination are taking measures to establish a collegiate institution in that State.

LOUISIANA—Has a college at Jackson, in the eastern part of the State. The Roman Catholics have a college at New Orleans.

ALABAMA.—There is a respectable collegiate institution, under the fostering care of the Methodist Episcopal church, at La Grange, in the north-western part of this State.

Academies have been established in various parts of the West, for both sexes, and there are female seminaries of character and standing at Pittsburgh, Cincinnati, Granville, Louisville, Lexington, Nashville, and many other places. Several more colleges, and a large number of minor institutions, will be needed very shortly to supply the demands for education in the West. The public mind is awake to the subject of education, and much has already been done, though a greater work has yet to be accomplished to supply the wants of the West in literary institutions.

An annual convention is held in Cincinnati, on the first Monday in October, denominated the "*Western Institute and College of Professional Teachers.*" Its object, according to the constitution, is, "to promote by every laudable means, the diffusion of knowledge in regard to education, and especially by aiming at the elevation of the character of teachers, who shall have adopted instruction as their regular profession." The first meeting was held in 1831, under the auspices of the "Academic Institute," a previously existing institution, but of more limited operations. The second convention, in 1832, framed a constitution and chose officers, since which time regular meetings have been held by delegates or individuals from various parts of the West, and a volume of Transactions of three or four hundred pages published annually.

II. THEOLOGICAL INSTITUTIONS.

The *Western Theological Seminary* at Alleghany town, opposite Pittsburgh, is under the jurisdiction of the general assembly of the Presbyterian church. It commenced operations in 1829. At *Canonsburg* is a seminary belonging to the Associate church, of which Dr. Ramsey is professor. The Associate Reformed church have a theological school in Pittsburgh, under charge of the Rev. John T. Pressly, D. D. The Baptist denomination are now engaged in establishing a manual labor academy in the vicinity of Pittsburgh, for both ministerial and general education.

The theological departments of Oberlin, Granville, and other collegiate institutions, have already been noticed. *Lane Seminary*, near Cincinnati, was founded in 1830, by

Messrs. E. & W. A. Lane, merchants, of New Orleans, who made a very liberal offer of aid. Its location is excellent, two and a half miles from Cincinnati, at Walnut Hills, and is under the charge of the Rev. Dr. Beecher, and a body of professors. Number of students about forty. The *Hanover Institution* in Indiana, has been noticed already. In the theological department are three professors and twelve students. The Baptists in this State are about establishing a manual labor seminary for ministerial and general education.

A valuable property has been purchased, adjoining Covington, Ky., opposite Cincinnati, and measures have been put in train to found a theological seminary by the Baptist denomination. The executive committee of the "*Western Baptist Education Society*," have this object in charge. The "*Alton Theological Seminary*," located at Upper Alton, Illinois, is under an organization distinct from that of *Shurtleff College*, already noticed. This institution has fifty acres of valuable land, and a stone edifice of respectable size, occupied at present in joint concern with the college, and a valuable library of several hundred volumes. Its organization has been but recently effected. Rev. L. Colby is professor, with eight students. Other institutions, having theological education, either in whole or in part, their object, are in contemplation.

Two remarks, by way of explanation, are

here necessary. 1. Most of the colleges and theological schools of the Western Valley, have facilities for manual labor, or are making that provision. In several, some of the students pay half, and even the whole of their expenses, by their own efforts. Public sentiment is awake to this subject, and is gaining ground. 2. In enumerating the students, the members of the preparatory departments are included, many of whom do not expect to pass through a regular collegiate course. The circumstances and wants of the country, from its rapid growth, seem to require the appendage of a large preparatory department to every college.

It may be well, to observe here, that a great and increasing demand exists in all the Western States, and especially those bordering on the Mississippi, for teachers of primary schools. Hundreds and thousands of moral, intelligent and pious persons, male and female, would meet with encouragement and success in this department of labor. It is altogether unnecessary for such persons to write to their friends, to make inquiries whether there are openings, &c. If they come from the older States, with the proper recommendations as to character and qualifications, they will not fail to meet with employment in almost any quarter to which they may direct their course. There is not a county in Missouri, Arkansas, Illinois, or Indiana, where persons would not meet with constant employment in teaching, and especially where teachers in Sabbath schools are needed. Persons desirous of such a field of humble, yet useful labor, should come here with the fixed purpose to mix with, and conform to the usages of the western population, to avoid fastidiousness, and to submit to the plain, frank, social and hospitable manners of the people.

III. DEAF AND DUMB ASYLUMS.

There are two institutions of this description in the West: one at Columbus, Ohio; the other at Danville, Ky. The one in Ohio contains about fifty pupils.

IV. MEDICAL INSTITUTIONS.

The medical department in Transylvania university, Kentucky, has six professors, and usually about two hundred students to attend the lectures. Fees for an entire course, with matriculation and library, one hundred and ten dollars. Two medical institutions of respectable standing exist in Cincinnati: one connected with the Miami university, the other with Cincinnati college.

The *Ohio Reformed Medical School* was established at Worthington, nine miles north of Columbus, in 1830. No specified time is required for study; but when a student will pass examination, he is licensed to practice.

V. LAW SCHOOLS.

The law department of Transylvania university is under the charge of two able professors, who hear recitations and deliver lectures. The average number of students is about forty.

A law school was established at Cincinnati, in 1833, with four professors:—Messrs. John C. Wright, John M. Goodenow, Edward King and Timothy Walker. The bar, the institution and the city have recently sustained a severe loss in the decease of Mr. King.

VI. BENEVOLENT AND RELIGIOUS SOCIETIES.

To enumerate and give particulars of all these, would make a volume. We can but barely call the attention of the reader to some of the more prominent organizations amongst the different Christian denominations in this great Valley, for doing good.

The *Foreign Missionary Society of the Valley of the Mississippi*, is a prominent auxiliary of the American Board of Commissioners for Foreign Missions. Its seat is Cincinnati; but by agencies and branches, it operates throughout the Valley. The Report of November, 1835, states, that *eighteen thousand six hundred and fifty-eight dollars* had been received into the treasury the preceding year. An edition of 3000 copies of the Missionary Herald is republished in Cincinnati, for circulation in the West.

The *Western Education Society*, connected with the American Education Society, has also its seat of operations at Cincinnati. Auxiliaries also exist in most of the Western States. Seventy-one beneficiaries were under its charge at the last anniversary.

The *American Tract Society* has auxiliaries and agencies in most of the Western States. The operations of the *American Bible Society*, through its numerous auxiliaries, is felt to the remotest parts of the West.

The *American Sunday School Union* has recently established a central agency in Cincinnati, and is preparing to renew and greatly enlarge its very important efforts for the benefit of the rising generation in the West.

A series of very interesting anniversaries are held in Cincinnati the first week in November, when all the great objects of Christian effort receive a renewed impulse.

The *American Home Missionary Society* has more than two hundred missionaries laboring in the States west of the mountains. In 1835, they assisted two hundred and seventeen Presbyterian ministers in this field.

The *Temperance effort* has not been neglected, and an interesting change is going forward, in a quiet and noiseless way, in the habits of the people, in reference to the use of intoxicating liquors. It is to be hoped, that more prompt and vigorous efforts will be made to promote this cause; but even now, there are many thousands, who abstain from the use of spirituous liquors without any formal pledge.

The *Methodist Episcopal church*, in addition to their regular system of circuits, are extending the influence of their denomination on the frontiers, by missionary operations, and their labors are prospered.

The *Baptist denomination* have made some important movements in the Western Valley within the last three years. Their Home Mission Society has nearly one hundred missionaries in the West. In November, 1833, the "*General Convention of Western Baptists*" was organized by more than one hundred ministers and brethren, assembled from various parts of the West. It is not an ecclesiastical body, claiming jurisdiction either over churches or ministers, nor is it strictly a missionary body. Its business, according to the constitution, is, "to promote, by all lawful means, the following objects, to wit:—Missions, both foreign and domestic; ministerial education, for such as may have first been licensed by the churches; Sunday schools, including Bible classes; religious periodicals; tract and temperance societies; as well as all others warranted by Christ in the gospel."

At its second session, in 1834, the "*Western Baptist Education Society*" was formed. Its object is "the education of those who give evidence to the churches of which they are members, that God designs them for the ministry." The executive committee are charged temporarily with establishing the Central Theological Seminary, already mentioned, at Covington, Ky.

Many other interesting associations for humane, philanthropic and religious purposes, exist in the Valley, which are necessarily omitted.

VII. THE PERIODICAL PRESS.

The number of different periodicals published in the Valley of the Mississippi must exceed four hundred, of which twelve or fifteen are daily papers. There are twenty-five weekly periodicals in Mississippi, one hundred and sixteen in Ohio, thirty-eight in Indiana, nineteen in Illinois, seventeen in Missouri, three, and probably more, in Arkansas, two, at least, in Wisconsin Territory. The *Western Monthly Magazine*, published at Cincinnati, is well known. The *Western Literary Journal and Monthly Review* is a respectable periodical, under the editorial management of W. D. Gallagher, Esq. The *Western Journal of the Medical and Physical Sciences*, edited by Daniel Drake, M. D., Professor of Theory and Practice of Medicine in the Cincinnati college, is published quarterly, in Cincinnati. There are a number of religious weekly, semi-monthly and monthly periodicals, devoted to the interests of the principal denominations through the Valley. There are known to be at least one in Western Virginia, two in Western Pennsylvania, seven in Ohio, four in Kentucky, four in Tennessee, two in Illinois, two in Missouri, and one in New Orleans. Supposing the average number of copies of western periodicals equalled seven hundred and fifty, this, estimating the different periodicals at four hundred, would give three hundred thousand! We see no marked and essential difference in the talent with which the editorial press is conducted, between the Eastern and Western States. The limits of this work will not allow me to add further evidence that our western population are not all "illiterate," and that "not more than one person in ten can read," than the following epitome of the issues of one of the publishing-houses in Cincinnati, as exhibited in the Cincinnati Journal.

"*Western Enterprise.* The enterprise of the West is not generally appreciated. As a specimen, we have procured from Messrs. Corey & Webster the following list of books, published by them within the last three years. These books are of sterling value.

Western Primer,	60,000
Webster's Spelling Book,	600,000
Primary Reader,	7,500
Elementary Reader,	37,000
Western "	16,000
Webster's History of the United States,	4,000
Miss Beecher's Geography,	15,000
Pocket Testament,	6,500
Watts' and Select Hymns,	8,000
Beecher's Lectures on Skepticism, (three editions, 1000 each),	3,000
Stowe's Introduction to the Study of the Bible,	1,500
Christian Lyre,	2,000
Mitchell's Chemistry,	1,000
Eberle on the Diseases of Children,	2,000
Eberle's Notes of Practice,	1,500
Young Lady's Assistant in Drawing,	1,000
Munsell's Map,	3,500
Chase's Statutes of Ohio, (three volumes,)	1,000
Hammond's Reports, (sixth volume,)	500
Total,	771,000

Probably some of the many other publishers in the city have got out nearly or quite as many books. Truly, we are a book-making and book-reading nation."

VIII. RELIGIOUS DENOMINATIONS.

In exhibiting the following statistics, entire correctness is not attempted. In some of the States, the latest reports have been had; in others, the author has taken data of two or three years date. Of the numbers of some of the numerous sects existing, the opinions of individuals have been the chief data he could obtain.

1. *Baptists.*

States and parts of States.	Churches.	Ministers.	Communicants.
Western Pennsylvania,	50	30	2,569
Western Virginia,	89	48	3,306
Ohio,	332	175	13,926
Michigan,	60	30	1,700
Indiana,	320	175	15,000
Illinois,	240	163	6,741
Missouri,	180	115	6,990
Arkansas,	25	18	700
Louisiana,	20	12	1,000
Mississippi,	100	46	4,000
North Alabama,	125	53	5,700
Tennessee,	348	292	22,868
Kentucky,	558	296	38,817
Total,	2447	1353	123317

Periodicals. The *Cross and Journal*, weekly, and *Baptist Advocate*, monthly, at Cincinnati;

the *Baptist Banner*, weekly, at Shelbyville, Ky.; the *Baptist*, a large monthly quarto, at Nashville, Tenn.; the *Western Pioneer and Baptist Standard-Bearer*, weekly, at Upper Alton, Ill.; and the *Witness*, a small quarto, weekly, at Pittsburgh.

2. *Methodists* (*Episcopal*). This denomination is divided into conferences, which are not arranged exactly with the boundaries of the States. A large book and printing-office is established at Cincinnati, where all the society's publications are kept for sale. Another depository is kept at Nashville.

Conferences.	Circuit Preachers, &c.	White Members.	Colored.	Indians.	Total number of members.
Mississippi, ...	55	6,358	2,622	727	9,707
Alabama, (one District in the Valley.)	16	3,051	490		3,543
Pittsburgh, ...	156	40,155	296		40,451
Ohio, ...	204	62,686	544	217	63,447
Missouri, (including Arkansas,)	57	7,948	1,061	889	9,898
Kentucky, ...	100	25,777	5,592		31,369
Illinois, ...	6	15,038	59		15,097
Indiana, ...	70	24,984	229		25,213
Holston, ...	62	21,559	2,478		24,031
Tennessee, ...	120	29,794	5,043	508	35,345
Total,	901	237,350	18,416	2,341	258,101

Allowing two *local* to one *circuit* preacher, which is rather under than over the proportion, would make 1802, which, added to the number of those whose names are on the minutes of the conferences, would make 2703 Methodist Episcopal ministers of the gospel in the Valley of the Mississippi. The *Pittsburgh Conference Journal*, *Western Christian Advocate*, and *Western Methodist*, are their periodicals.

3. *Methodists* (*Protestant*). There are two conferences of this denomination, in the West; the Pittsburgh and Ohio conferences and their circuits, preaching-stations and members, extend through the States north of the Ohio river, with a few stations and churches south.

Pittsburgh Conference has twenty-eight circuit and eighty-five local preachers and licentiates; twenty-five circuits, four stations and two mission-circuits, with six thousand nine hundred and two members in society.

Ohio Conference has twenty-eight circuit and ninety local preachers, twenty-two circuits, three stations, three mission-circuits, and three thousand six hundred and sixty-seven members. The *Methodist Correspondent*, a neat, semi-monthly quarto periodical, published at Zanesville, Ohio, is devoted to their interests.

4. *Presbyterians.* The following table (with the exception of Illinois), embraces fifty-six presbyteries, and is constructed from the returns to the general assembly, in 1834; the minutes of 1835, we understand, have not been printed.

States and parts of States.	Churches.	Ministers.	Communicants.
Western Pennsylvania and Western Virginia,	212	135	22,687
Michigan,	32	20	1,397
Ohio,	400	255	27,821
Indiana,	99	55	4,339
Illinois,	71	50	2,000
Missouri,	33	20	1,549
Arkansas,	12	9	390
Kentucky,	120	83	8,378
Tennessee,	121	90	9,926
North Alabama,	15	12	725
Mississippi,	33	24	761
Total,	1148	753	79,973

Periodicals. The *Cincinnati Journal and Western Luminary*, published at Cincinnati; *Christian Herald*, at Pittsburgh; *Ohio Observer*, at Hudson, Ohio; *Western Presbyterian Herald*, at Louisville, Ky.; *New Orleans Observer*, at New Orleans; and *Alton Observer*, at Alton, Ill.;—all weekly; and the *Missionary Herald*, republished at Cincinnati, monthly.

5. *Cumberland Presbyterians.* This sect originated from the Presbyterian church in 1804, in Kentucky, but did not increase much till 1810 or 12. They are spread through most of the Western States, and have thirty-four presbyteries, seven synods, and one general assembly. The minutes of their general assembly, now before me, are not sufficiently definite to give the number of congregations.

These probably exceed three hundred. An intelligent member of that denomination states the number of ordained preachers to be three hundred, licentiates, one hundred, candidates for the ministry, one hundred and fifty, and communicants, 50,000.

Periodicals. The *Cumberland Presbyterian* is a weekly paper, published at Nashville, Tenn. Another has been recently started at Pittsburgh.

6. *Congregationalists.* In Ohio, especially in the northern part, are a number of Congregational churches, and some ministers, as there are in Indiana, Michigan and Illinois. There are two or three ministers, twelve or fifteen congregations, and about five hundred communicants in Illinois, who are organized into an association in Illinois.

7. *Protestant Episcopal Church.* This denomination has seven dioceses in the Western or south-western States, exclusive of Western Pennsylvania and Western Virginia, which belong to the dioceses of those States. They are, Ohio, Michigan, Illinois, Kentucky, Tennessee, Mississippi, Indiana and Missouri. There are about seventy-five or eighty ministers, and twice as many churches in the West. Provision has been made, in part, for the endowment of the theological seminary at Gambier, O., in England, and Bishop McIlvaine has obtained about $12,600, to be appropriated in the erection of a gothic edifice, to be called "Bexley Hall," with three stories, and

accommodations for fifty students. A weekly periodical is issued at the same place, to support the interests of the denomination.

8. *German Lutherans.* We have no data to give the statistics of this denomination. There is a synod in Ohio, another in Western Pennsylvania, and perhaps others. There are probably fifty or sixty ministers in the West, and one hundred and fifty congregations.

9. *German Reformed Church.* There are eighty congregations in Ohio, twenty in Indiana, and probably fifty others in the West, with forty or fifty ministers.

10. The *Tunkers*, or *Dunkards*, have forty or fifty churches, and about half as many ministers in the Western States.

11. The *Shakers* have villages in several places in Ohio and Kentucky, but are losing ground.

12. The *Mormons* have a large community at Kirkland, Ohio, where, under the direction of their prophet, Joseph Smith, they are building a vast temple. They have probably two hundred preachers, and as many congregations in the West, and still make proselytes.

13. *Christian Sect*, or *Newlights*, have become, to a considerable extent, amalgamated with the "*Reformers*," or "*Campbellites.*" I have not data on which to construct a tabular view of this sect; but from general information, estimate the number of their "bishops" and "proclaimers" at three hundred, and their communicants at 10,000 or 12,000.

They have three or four monthly periodicals.

Alexander Campbell, who may be justly considered the leader of this sect (though they disclaim the term *sect*), is a learned, talented, and voluminous writer. He conducts their leading periodical,—the *Millennial Harbinger*.

14. The "*United Brethren in Christ*," are a pious, moral and exemplary sect, chiefly in Ohio, but scattered somewhat in other Western States. They are mostly of German descent, and, in their doctrinal principles and usages, very much resemble the Methodists. They have about three hundred ministers in the West, and publish the *Religious Telescope*, a large weekly paper, of evangelical principles, and well conducted,—printed at Circleville, Ohio.

15. *Reformed Presbyterians*, or *Covenanters*, have twenty or thirty churches, and as many ministers, but are much dispersed through the Northern Valley.

16. The *Associate Church*, or *Seceders*, are more numerous than the Covenanters.

17 The *Associate Reformed Church*. The western synod of this body still exists as a separate denomination. Their theological school, at Pittsburgh, has already been noticed. I know not their numbers, but suppose they exceed considerably the Associate Church.

18. The *Friends*, or *Quakers*, have a num-

ber of societies in Western Pennsylvania, Ohio, Indiana, &c.

19. The *Unitarians* have societies and ministers at Pittsburgh, Cincinnati, Louisville, St. Louis, and probably in other places.

There are many other sects and fragments in the West. The Valley of the Mississippi, like all new countries, is a wide and fertile field for the propagation of error, as it is for the display of truth.

20. *Roman Catholics.* The number of papal dioceses in the Valley, including the one at Mobile, is *seven*, of each of which a very brief sketch will be given, commencing with,

1. *Detroit*, including Michigan and the North-Western Territory,—one bishop, with sub-officers, eighteen priests, and as many chapels. At Detroit and vicinity, for two or three miles, including the French, Irish and Germans, Roman Catholic families make up one third of the population; probably 3,500, of all ages. At Ann Arbor, and in the towns of Webster, Scio, Northfield, Lima and Dexter are many. At and near Bertrand on the St. Joseph's river, adjoining Indiana, they have a school established and an Indian mission. Including the fur traders and Indians, they may be estimated at 10,000 in this diocess.*

* The reader will note that our estimates of Roman Catholics include the whole family of every age. Whereas, our statistics of Protestant denominations included only communicants.

2. *Cincinnati.* A large cathedral has been built in this place, and fifteen or twenty chapels in the diocess. Ten years ago, the late bishop Fenwick could not count up five hundred. The emigration of foreigners, and the laborers on the Ohio canals, and not a little success in proselyting, account for the increase. There are twenty-five congregations, and eighteen priests. A literary institution, called the *Athenæum*, is established at Cincinnati, where the students are required to attend the forms of worship, and the superior inspects all their letters. St. Peter's Orphan Asylum, is under charge of four "Sisters of Charity." The number of Catholics in Cincinnati is variously estimated, the medium of which is 6000, and as many more dispersed through the State.

3. *Bardstown.* This includes the State of Kentucky, and has a bishop, with the usual subordinates, twenty-seven congregations, and thiry-three priests, eleven of whom reside at Bardstown. A convent of six Jesuit priests at Lebanon; another of five Dominicans, called St. Rose, in Washington county; the college at Bardstown, already noticed, and St. Mary's Seminary in Washington county, for the education of priests. Of *female* institutions, there are the *Female Academy of Nazareth*, at Bardstown, conducted by the "Sisters of Charity," and superintended by the bishop and professors of St. Joseph's college,—one hundred and fifty pupils; the

female academy of Loretto, Washington county, with accommodation for one hundred boarders, and directed by the "*Sisters of Mary at the foot of the cross.*" This order have six other places for country schools, and are said to be one hundred and thirty-five in number. The *Convent of Holy Mary*, and the *Monastery of St. Magdalene*, at St. Rose, Washington county, by Dominican nuns, fifteen in number, and in 1831, thirty pupils. The Catholics have a female academy at Lexington, with one hundred pupils.

I have no data to show the Roman Catholic population of this State, but it is by no means proportionate to the formidable machinery here exhibited. All this array of colleges, seminaries, monasteries, convents and nunneries is for the work of proselyting, and if they are not successful, it only shows that the current of popular sentiment sets strongly in another direction.

4. *Vincennes.* This is a new diocess, recently carved out of Indiana and Illinois, by the authority of an old gentleman, who lives in the city of Rome! It includes a dozen chapels, four or five priests, the St. Claire convent at Vincennes, with several other appendages. The Roman Catholic population of this State is not numerous, probably not exceeding 3000. Illinois has about 5000, a part of which is under the jurisdiction of St. Louis diocess. In Illinois, there are ten churches and six priests, a part of which are

included in the diocess of Indiana. A convent of nuns of the "*Visitation of the blessed Virgin Mary*," at Kaskaskia, who conduct a female school, with a few boarders and about thirty or forty day scholars.

5. *St. Louis.* This diocess includes eighteen congregations and nineteen priests, with the following appendages:—1. *St. Louis University*, already noticed, with six priests, for instructers, and one hundred and fifty students, of which, about eighty are boarders. The rules require their attendance on morning and evening prayers, the catechism, and divine service on Sundays and holidays. 2. St. Mary's college, also noticed in our description of colleges. 3. Noviciate for *Jesuits under St. Stanislaus*, in St. Louis county.

Of female institutions there are,—1. Convent of the "*Ladies of the Sacred Heart*," at St. Louis: 2. another of the same description, and their noviciate, at Florrissant; 3. another of the same order at St. Charles; 4. a female academy at Carondalet, six miles below St. Louis, by the "*Sisters of Charity*;" 5. a convent and academy of the "*Sisters of Loretto*," at New Madrid; 6. a convent and female academy at Frederickstown, under supervision of a priest; 7. a convent and female academy of the "*Sisters of Loretto*," in Perry county. The Roman Catholic population in Missouri does not exceed 15,000. Their pupils of both sexes, may be estimated at seven hundred. To the above may be added the hospital, and

16*

the asylum for boys, in St. Louis, under the management of the Sisters of Charity.

Roman Catholic teachers, usually foreigners, disperse themselves through the country, and engage in teaching primary schools; availing themselves of intercourse with the families of their employers to instruct them in the dogmas of their religion. The greatest success that has attended the efforts of the priests in converting others, has been during the prevalence of the cholera, and especially after collapse and insensibility had seized the person! We know of more than sixty Roman Catholics who have been converted to the faith of Christ, and joined Christian churches within three or four years past, in this State.

6. *New Orleans.*—The Roman Catholics in Louisiana are numerous, probably including one third of the population. Relatively, Protestants are increasing, as a large proportion of the emigration from the other States, who care any thing about religion, are Protestants. There are twenty-six congregations, and twenty-seven priests with several convents, female seminaries, asylums, &c.

7. *Mobile.*—A splendid cathedral has been commenced here. This diocess extends into Florida.

CHAPTER XVII.

SUGGESTIONS TO EMIGRANTS.

Canal, Steam-Boat and Stage Routes — Other Modes of Travel — Expenses — Roads, Distances, &c., &c.

In the concluding chapter to this Guide, it is proposed to give such information as is always desirable to emigrants upon removing, or traveling for any purpose, to the West.

1. Persons in moderate circumstances, or who would save time and expense, need not make a visit to the West, to ascertain particulars previous to removal. A few general facts, easily collected from a hundred sources, will enable persons to decide the great question, whether they will emigrate to the Valley. By the same means, emigrants may determine to what State, and to what part of that State, their course shall be directed. There are many things that a person of plain, common sense will take for granted without inquiry,— such as facilities for obtaining all the necessaries of life; the readiness with which property of any description may be obtained for a fair

value, and especially farms and wild land; that they can live where hundreds of thousands of others of similar habits and feelings live; and above all, they should take it for granted, that there are difficulties to be encountered in every country, and in all business;—that these difficulties can be surmounted with reasonable effort, patience and perseverance, and that in every country, people sicken and die.

2. Having decided to what State and part of the State, an emigrant will remove, let him then conclude to take as little furniture and other luggage as he can do with, especially if he comes by public conveyances. Those who reside within convenient distance of a sea port, would find it both safe and economical to ship by New Orleans, in boxes, such articles as are not wanted on the road, especially if they steer for the navigable waters of the Mississippi. Bed and other clothing, books, &c., packed in boxes, like merchants' goods, will go much safer and cheaper by New Orleans, than by any of the inland routes. I have received more than one hundred packages and boxes from eastern ports, by that route, within twenty years, and never lost one. Boxes should be marked to the owner or his agent at the river port where destined, and to the charge of some forwarding house in New Orleans. The freight and charges may be paid when the boxes are received.

3. If a person designs to remove to the north part of Ohio and Indiana, to Chicago and vicinity, or to Michigan, or Green Bay, his course should be by the New York canal, and the lakes. The following table, showing the time of the opening of the canal at Albany and Buffalo, and the opening of the lake, from 1827 to 1835, is from a report of a committee at Buffalo to the common council of that city. It will be of use to those who wish to take the northern route in the spring.

Year.	Canal opened at Buffalo.	Canal opened at Albany.	Lake Erie opened at Buffalo.
1827	April 21	April 21	April 21
1828	" 1	" 1	" 1
1829	" 25	" 29	May 10
1830	" 15	" 20	April 6
1831	" 16	" 16	May 8
1832	" 18	" 25	April 27
1833	" 22	" 22	" 23
1834	" 16	" 17	" 6
1835	" 15	" 15	May 8

The same route will carry emigrants to Cleaveland, and by the Ohio canal, to Columbus, or to the Ohio river, at Portsmouth; from whence, by steam-boat, direct communications will offer to any river port in the Western States. From Buffalo, steam-boats run constantly (when the lake is open), to Detroit, stopping at Erie, Ashtabula, Cleaveland, Sandusky and many other ports, from whence stages run to every prominent town. Transportation-wagons are employed in forwarding goods.

Route from Buffalo to Detroit, by water.

	Miles.		Miles.
Dunkirk, N.Y.	39	Cleaveland, Ohio, .	30—193
Portland, "	18—57	Sandusky, " .	54—247
Erie, Pa.	35—92	Amherstburg, N.C.	52—299
Ashtabula, Ohio, .	39—131	Detroit, Mich., . . .	18—317
Fairport, " .	32—163		

From Detroit to Chicago, Illinois.

	Miles.		Miles.
St. Clair river, Mich., .	40	Mackinaw, ...	58—329
Palmer, ...	17—57	Isle Brule, ...	75—404
Fort Gratiot, ...	14—71	Fort Howard, Wisconsin Ter., . . .	100—504
White Rock, ...	40—111	Milwaukee, W.T.	310—814
Thunder Island, . .	70—181	Chicago, Ill., . . .	90—904
Middle Island, . . .	25—206		
Presque Isle, ...	65—271		

Fron Cleaveland to Portsmouth, via. the Ohio canal.

	Miles.		Miles.
Cuyahoga aqueduct,	22	Irville, ...	26—158
Old Portage, ...	12—34	Newark, ...	13—171
Akron, ...	4—38	Hebron, ...	10—181
New Portage, ...	5—43	Licking Summit, . .	5—186
Clinton, ...	11—54	Lancaster Canaan,	11—197
Massillon, ...	11—65	Columbus, side cut,	18—215
Bethlehem, ...	6—71	Bloomfield, ...	8—223
Bolivar, ...	8—79	Circleville, ...	9—232
Zoar, ...	3—82	Chillicothe, ...	23—255
Dover, ...	7—89	Piketon, ...	25—280
New Philadelphia, .	4—93	Lucasville, ...	14—294
Newcomers'town,	22—115	Portsmouth, (Ohio river), ...	13—307
Coshocton, ...	17—132		

The most expeditious, pleasant and direct route for travelers to the southern parts of Ohio and Indiana; to the Illinois river, as far north as Peoria; to the Upper Mississippi, as far as Quincy, Rock island, Galena and Prairie du Chien; to Missouri, and to Kentucky, Tennessee, Arkansas, Natchez and New Orleans, is one of the southern routes. These are,—1. From Philadelphia to Pittsburgh, by rail-roads and the Pennsylvania canal; 2. By the Baltimore and Ohio rail-road and stages, to Wheeling; or, 3. For people living to the south of Washington, by stage, by the way of Charlottesville, (Virginia,) Staunton, the Hot, Warm, and White-Sulphur Springs, Lewisburg, Charlestown, to Guyandotte, from whence a regular line of steam-boats runs three times a week to Cincinnati. Intermediate routes from Washington city to Wheeling, or to Harper's Ferry, to Fredericksburg, and intersect the route through Virginia, at Charlottesville.

From Philadelphia to Pittsburgh, by the rail-road and canal.

	Miles.
Columbia, on the Susquehanna river, by rail-road, daily, ...	81
By canal packets, to	
Bainbridge, ...	11—92
Middletown, ...	17—109
Harrisburg, ...	10—119
Juniata river, ...	15—144
Millerstown, ...	17—151
Mifflin, ...	17—168
Lewistown, ...	13—181
Waynesburg, ...	14—195
Hamiltonville, ...	11—206
Huntingdon, ...	7—213
Petersburg, ...	8—221
Alexandria, ...	23—244
Frankstown and Hollidaysburg, ...	3—247
From thence, by rail-road, across the mountain, to	
Johnstown, ...	38—285
By canal, to	
Blairsville, ...	35—320
Saltzburg, ...	18—338
Warren, ...	12—350
Alleghany river, ..	16—366
Pittsburgh, ...	28—394

The *Pioneer* line, on this route, is exclusively for passengers, and professes to reach Pittsburgh in four days, but is sometimes behind, several hours. Fare through, $10. Passengers pay for meals.

The *Good-Intent* line is also for passengers only, and runs in competition with the Pioneer line.

Leech's line, called the "*Western Transportation line*," takes both freight and passengers. The packet-boats advertise to go through, to Pittsburgh, in five days, for $7.

Midship and steerage passengers in the transportation line, in six and a half days,—merchandise delivered in eight days. Generally, however, there is some delay. Emigrants must not expect to carry more than a small trunk or two, on the packet-lines. Those who take goods or furniture, and wish to keep with it, had better take the transportation lines, with more delay. The price of meals on board the boats is about thirty-seven and a half cents.

In all the steam-boats on the Western waters no additional charge is made to cabin passengers for meals;—and the tables are usually profusely supplied. Strict order is observed, and the waiters and officers are attentive.

Steam-boat route from Pittsburgh to the mouth of the Ohio.

	Miles.		Miles.
Middletown, Pa. . . .	11	Steubenville, Ohio, .	27—69
Economy, " . . .	8—19	Wellsburgh, Va., . .	7—76
Beaver " . . .	10—29	Warren, Ohio, ...	6—82
Georgetown, " . . .	13—42	Wheeling, Va., . . .	10—92

	Miles.
Elizabethtown, Va.,	11—103
Sistersville, "	34—137
Newport, Ohio,	27—164
Marietta, "	14—178
Parkersburgh, Va.	11—189
Belpre and Blannerhasset Island, O.,	4—193
Troy, Ohio,	10—203
Belleville, Va.,	7—210
Letart's Rapids, "	37—247
Point Pleasant, "	27—274
Gallipolis, Ohio,	4—278
Guyandotte, Va.,	27—305
Burlington, Ohio,	10—315
Greensburg, Ky.,	19—334
Concord, Ohio,	12—346
Portsmouth(Ohio canal),	7—353
Vanceburg, Ky.,	20—373
Manchester, Ohio,	16—389
Maysville, Ky.,	11—400
Charleston, "	4—404
Ripley, Ohio,	6—410
Augusta, Ky.,	8—418
Neville, Ohio,	7—425
Moscow, "	7—432
Point Pleasant, "	4—436
New Richmond, "	7—443
Columbia, "	15—458
Fulton, "	6—564
CINCINNATI, "	2—466
North Bend, "	15—481
Lawrenceburgh, Ind. and mouth of the Miami,	8—489
Aurora, Ind.,	2—491
Petersburg, Ky.,	2—493
Bellevue, "	8—501
Rising Sun, Ind.,	2—503
Fredericksburg, Ky.	18—521

	Miles.
Vevay, Ind., and Ghent, Ky.,	11—532
Port William, Ky.,	8—540
Madison, Ind.,	15—555
New London, "	12—567
Bethlehem, "	8—575
Westport, Ky.,	7—582
Transylvania, "	15—595
LOUISVILLE, "	12—609
Shippingport, through the canal,	2½—611½
New Albany, Ind.,	1½—613
Salt River, Ky.,	23—636
Northampton, Ind.,	18—654
Leavenworth, "	17—671
Fredonia, "	2—673
Rome, "	32—705
Troy, "	25—730
Rockport, "	16—746
Owenburgh, Ky.,	12—758
Evansville, Ind.,	36—794
Henderson, Ky.,	12—806
Mount Vernon, Ind.	28—834
Carthage, Ky.,	12—846
Wabash river, "	7—853
Shawneetown, Ill.,	11—864
Mouth of Saline, "	12—876
Cave in Rock, "	10—886
Golconda, "	19—905
Smithland, mouth of the Cumberland river, Ky.,	10—915
Paducah, mouth of the Tennessee river, Ky.,	13—928
Caledonia, Ill.,	31—959
Trinity, mouth of Cash river, Ill.,	10—969
MOUTH OF THE OHIO RIVER,	6—975

Persons who wish to visit Indianopolis will stop at Madison, Indiana, and take the stage conveyance. From Louisville, by the way of Vincennes, to St. Louis by stage, every alternate day, two hundred and seventy-three miles, through in three days and a half. Fare $17. Stages run from Vincennes to Terre Haute and other towns up the Wabash river. At *Evansville*, Indiana, stage lines are connected with Vincennes and Terre Haute; and at *Shawneetown* twice a week to Carlyle, Illinois, where it intersects the line from Louisville to St. Louis. From Louisville to Nashville by steam-boats, passengers land at Smithland at the mouth of Cumberland river, unless they embark direct for Nashville.

In the *winter*, both stage and steam-boat lines are uncertain and irregular. Ice in the rivers frequently obstructs navigation, and high waters and bad roads sometimes prevent stages from running regularly.

Farmers who remove to the West from the Northern and Middle States, will find it advantageous, in many instances, to remove with their own teams and wagons. These they will need on their arrival. Autumn, or from September till November, is the favorable season for this mode of emigration. The roads are then in good order, the weather usually favorable, and feed plenty. People of all classes, from the States south of the Ohio river, remove with large wagons, carry and cook their own provisions, purchase their feed by the bushel, and invariably *encamp out at night.*

Individuals who wish to travel through the interior of Michigan, Indiana, Illinois, Missouri, &c., will find that the most convenient, sure, economical and independent mode is on horseback. Their expenses will be from seventy-five cents to one dollar fifty cents per day, and they can always consult their own convenience and pleasure as to time and place.

Stage fare is usually six cents per mile, in the West. Meals, at stage-houses, are thirty-seven and a half cents.

Steam-boat Fare, including Meals.

From Pittsburgh to Cincinnati,	$10
" Cincinnati to Louisville,	4
" Louisville to St. Louis,	12

And frequently the same from Cincinnati to St. Louis,—varying a little, however.

A *deck* passage, as it is called, may be rated as follows:—

From Pittsburgh to Cincinnati,	$3
" Cincinnati to Louisville,	1
" Louisville to St. Louis,	4

The deck for such passengers is usually in the midship, forward of the engine, and is protected from the weather. Passengers furnish their own provisions and bedding. They often take their meals at the cabin table, with the boat-hands, and pay twenty-five cents a meal. Thousands pass up and down the rivers as deck passengers, especially emigrating families, who have their bedding, provisions and cooking utensils on board.

The whole expense of a single person from New York to St. Louis, by the way of Philadelphia and Pittsburgh, with cabin passage on the river, will range between $40 and $45;—time from twelve to fifteen days.

Taking the transportation lines on the Pennsylvania canal, and a deck passage in the steam-boat, and the expenses will range between $20 and $25, supposing the person buys his meals at twenty-five cents, and eats twice a day. If he carry his own provisions, the passage, &c., will be from $15 to $18.

The following is from an advertisement of the Western Transportation, or Leech's line, from Philadelphia:—

	Miles.	Days.	Fare.
Fare to Pittsburgh,	400	6½	$6 00
" Cincinnati,	900	8½	8 50
" Louisville,	1050	9½	9 00
" Nashville,	1650	13½	13 00
" St. Louis,	1750	14	13 00

The above does not include meals.

Packet-boats for Cabin Passengers (same line.)

	Miles.	Days.	Fare.
Fare to Pittsburgh,	400	5	$7 00
" Cincinnati,	900	8	17 00
" Louisville,	1050	9	19 00
" Nashville,	1650	13	27 00
" St. Louis,	1750	13	27 00

Emigrants and travelers will find it to their interest always to be a little skeptical relative to statements of stage, steam and canal-boat agents, to make some allowance in their own calculations for delays, difficulties and expenses and above all, to *feel* perfectly patient and in good humor with themselves, the officers, company, and the world, even if they do not move quite as rapid and fare quite as well as they desire.

Spotted something that isn't quite right?

Let us know and we'll fix it for future reprints. And as a thank you for your diligence, we'll send you a free Sothis book of your choice.

publisher@sothis.press

This page intentionally left blank

Except of course for that line above. And this one.
And whilst we have your attention, perhaps you'd
like to follow us on Instagram @sothis.press. Go on.

Made in United States
Cleveland, OH
06 November 2026

47795285R00227